FEATURES OF EVOLUTION in the FLOWERING PLANTS

By Ronald Good

M.A., Sc.D. (Cantab.)

Emeritus Professor of Botany, The University of Hull

ILLUSTRATED BY

Marjorie E. Cunningham

B.Sc. (Leeds)

and the author

DOVER PUBLICATIONS, INC.

NEW YORK

TO MY WIFE

ACKNOWLEDGMENTS

Figs. 18 and 58 are modified from Brown, *The Plant Kingdom*, by permission of Messrs. Ginn & Co., Boston, Mass. and Figs. 12, 14, 23, 29, 37, 39, 44, 73 and 96 are modified from Bailey, The Standard Cyclopedia of Horticulture by permission of The Macmillan Company, New York.

Published in Canada by General Publishing Company, Ltd., 30 Lesmill Road, Don Mills, Toronto, Ontario.
Published in the United Kingdom by Constable and Company, Ltd., 10 Orange Street, London WC 2.

This Dover edition, first published in 1974, is an unabridged and slightly corrected republication of the work originally published by Longmans Green & Co., Ltd., in London, in 1956. The author has written a new Preface especially for the Dover edition.

International Standard Book Number: 0-486-61591-X
Library of Congress Catalog Card Number: 74-82210

Manufactured in the United States of America
Dover Publications, Inc.
180 Varick Street
New York, N.Y. 10014

PREFACE TO THE FIRST EDITION

WHATEVER opinion may be held about the idea of evolution itself, or about its possible modes of operation, it cannot be denied that change with time is an all-pervading principle of the natural world. Nor can it be denied that both our knowledge of the facts of nature and our appreciation of them, are as much subject to this principle as anything else. In consequence of it understanding grows and theory changes but it is not always easy to relate the two and we must always be alert to the necessity of keeping them in step. It is now nearly a hundred years since the Darwinian theory of evolution became current, but neither the circumstances nor the theories of biology are now those of the eighteen-fifties: there has been change in both, and there are some today who doubt whether the two have fully kept in phase with one another. The need for a fresh assessment more in keeping with the knowledge of the times has, indeed, become more and more pressing. Many particular problems require to be thought out again from the beginning, but the special need is for a new, objective, and sober consideration of the facts, and, above all, of the facts in some of those aspects of biology which have so far received less attention, and which are therefore less obscured by the patina of controversy. It is as a modest contribution to such a new judgment that this book, which embodies the thoughts and observations of many years, has been written, and its purpose should not, through any default or omission on the part of its author, be left open to misunderstanding. Its primary aim is to redirect attention to facts concerning one great and fundamentally important section of the living world which, at best, have become forgotten in the pursuit of other and more fashionable lines of enquiry, or, at worst, have never been fully realized at all. Its second purpose is to demonstrate that when these neglected facts are taken properly into account, some, at least, of the best-known speculations about organic evolution are seen to have a less general applicability than is usually claimed. Lastly, the book tries to summarize the particular conclusions to which a judicial consideration of the facts that it presents appears to lead.

Because the book is intended to be a step towards a fresh and untrammelled outlook on some of the problems of evolution, it has been felt unnecessary, and indeed undesirable, to burden it with many references to what, with the best intentions in the world, could only be an invidious and meagre selection from the vast general literature of that subject. Instead, there have been appended to appropriate chapters short notes giving a few chosen sources from which the reader should be able to pursue their respective subjects rather further and, especially, to find more widely ranging bibliographies.

Many of the plants mentioned are likely to be relatively unfamiliar to most readers and illustrations of them are therefore unusually important. This point has been met partly by providing as many figures as possible, and partly by selecting for mention, whenever feasible, plants of which there are illustrations in the first or second editions of Engler and Prantl's *Die Naturlichen Pflanzenfamilien* or in Baillon's *Histoire des Plantes* (and its English translation). For the rest sources of pictures should be sought in Stapf's *Iconum Botanicarum Index Londinensis*. The numbers of genera and species given, which are intended only for purposes of broad comparison, are based on those in Lemée's *Dictionnaire descriptif et synonymique des genres de plantes phanérogames*.

For reasons which will be sufficiently apparent as the reader proceeds plant groups are referred to by familiar or anglicized names whenever this can be done without too much ambiguity, but these claim no more than the virtue of immediate convenience. The word *taxon*, with its plural *taxa*, is used sometimes when the employment of a more precise categorical term is undesirable. Many of the lists of examples quoted from time to time are, it is believed, reasonably full, but it must not be assumed that any one of them is entirely complete.

The chapters, apart from the first and last, fall into three groups. The second and third form what may be looked upon as a background to the Flowering Plants; the next three provide a review of that whole group, written in somewhat novel terms and with appropriate reference to the evolutionary problems which it presents as a whole, and the remaining five cover a small selection of the many more particular problems which make these plants of special interest in the study of evolution. It may be noticed that one of the most obvious of these, namely the "insect-flower relation", is not dealt with, the reason being that it is so profound a subject and involves so many aspects of biology that it could not have been treated adequately in the space available here.

For the views expressed in this book I, alone, am responsible, but on matters of fact I have received during the writing of it much generous help and advice from many friends and I would ask them all to accept my thanks.

I am most grateful to Miss Malins for the long series of beautiful drawings by which Chapters 5 and 6, in particular, are enhanced. They were, perforce, made under conditions far from ideal, and among many other pressing demands on her time, and I hope that she may feel rewarded by the notable contribution that they make to the book.

Hull, *September*, 1955.   RONALD GOOD.

PREFACE TO THE DOVER EDITION

SINCE the first edition of this book was published, various other plants which might have been included in the appropriate places have come to my notice, and a few of these are sufficiently remarkable to call for mention here.

Dichromena, belonging to the Cyperaceae, should certainly be added to the examples of involucration given in the second paragraph of p. 288, and the Euphorbieae deserve much fuller treatment than they receive on pp. 290 and 291. Indeed, a thorough study of this group would make an admirable subject for research. Some of its members especially invite further investigation, notably *Euphorbia leucocephala* which, when in flower, has all the appearance of a characteristic member of the Umbelliferae; and the extraordinary *Neoguillauminia cleopatra,* of New Caledonia, which has a particularly complicated kind of pseudanthium which may well be unique.

In 1966 there was published the long-awaited seventh edition of *A Dictionary of the Flowering Plants and Ferns* by the late Dr. J. C. Willis (edited by H. K. Airy Shaw, Cambridge University Press); this work contains much of the information in Lemée's *Dictionnaire descriptif et synonymique des genres de plantes phanérogames* in more up-to-date form, and is much more concise and easily handled. Some of the details and figures in it differ from those quoted in the following pages, but this is accounted for partly by the lapse of time and partly by differences of personal opinion and does not affect the general argument of the book in any way.

Parkstone, 1974.   RONALD GOOD.

CONTENTS

ILLUSTRATIONS

CHAPTER ONE

INTRODUCTION

FOR nearly a century now a belief that the many thousands of different kinds of plants and animals at present inhabiting the earth have come into being by descent from fewer and earlier kinds through a process of continuing gradual change, that is to say by what is commonly described as a process of organic evolution, has been a fundamental article of biological faith. Indeed, it may be regarded as the most fundamental of all such articles, for it has become the basic organizing concept of the science of living things, so that to-day there rests upon it, either indirectly or directly, almost the whole fabric of modern thought. All sorts of people, by no means all of them biologists, have been nurtured in it and influenced by it; it permeates biological research and experiment; and there has grown up about it a vast literature.

All this being so it might be expected that there would be an essential unanimity of opinion about the probable means by which evolution has been effected and that writings on the subject would be distinguished by the highest standards of scientific dialectic, but in neither case can this be claimed. There are still deep divergences of opinion about the nature of evolutionary processes, while it is the experience of many that increased acquaintance with the literature brings more rather than less uncertainty, together with a conviction that however much may already have been said there is even more still to be learned. Moreover, much of this literature has something of the alluring but elusive quality of a mirage, in which the scene, at first apparently so sharply etched, gradually dissolves as it is more closely approached until it loses much of its earlier certainty of outline.

In consequence, although evolution finds wide tacit acceptance as a grand organizing concept of biology and as the best available working hypothesis to explain the present multifariousness of plant and animal life, many people gravely doubt the validity of many of the more particular arguments by which it is customarily sustained.

Some even question the whole idea. The biological sciences today are thus in the uneasy position of having to use, as one of their principal tools, a body of theory which is both sententious and incomplete, and one in which many workers have less than complete confidence.

The factors which have brought about this situation are many, and it is important to realize what some at least of them are. The more immediate cause is that attention has been unduly concentrated on certain parts only of the living world, so that our ideas and conclusions about evolution rest on too uneven a knowledge of all the relevant facts. Some aspects of the subject have been pursued intensively; others have received little more than perfunctory notice; while others again have attracted barely any attention at all, and this variation bears little relation to the intrinsic importance of the aspects concerned. As a result the whole approach to evolutionary problems lacks balance because it is based too much on argument from the particular to the general. There has been too much induction and too little deduction.

The more antecedent causes are broadly three, namely, difficulties inherent in the direct study of evolution; historical circumstances; and certain more subjective influences.

The fundamental inherent difficulty in the study of evolution is that this great natural process involves time dimensions of a magnitude quite out of proportion to the duration of human life or even to the sum of human experience, and the observer has therefore to rely on indirect, or circumstantial evidence. Hence beliefs that are often referred to as *theories* of evolution are, more accurately, only working hypotheses. This is a very important matter because the essence of a hypothesis is that it is an opinion suggested by the available evidence, but not one which precludes the possibility of some alternative. A hypothesis may well be substantiated when more corroborative details are forthcoming, but until then there is no logical reason for excluding the consideration of some other explanation of the facts. So, while it may be justifiable to believe that evolution affords a reasonable explanation of the facts of nature, it is not justifiable to maintain that no other explanation is possible or permissible.

The great difference between direct and indirect evidence is that the latter is cumulative. In direct evidence the quantity available (provided of course that it is not mistaken evidence) is relatively unimportant, because what is once shown to be true needs no further demonstration. In circumstantial, or indirect, evidence however, the greater the amount, and even more the greater the variety, of such evidence, and the more closely it relates to the question at issue, the stronger becomes the likelihood of the conclusion to which it points. The probability of the truth of any proposition is therefore greater or less according to the quantity and quality of the circumstantial evidence relating to it. But evidence of this kind cannot supply a final proof; it cannot amount to more than an extremely high degree of probability.

From the earliest days this dependence on circumstantial evidence went far to determine the directions that evolutionary enquiry would take, and the main lines of this evidence, which is very strong, are well known, being what the text-books call the "evidences for evolution", so that there is no need to make more than a brief reference to them here.

First, there are those lines which derive their value from the support they give to the opinion that existing organisms are mutually related in the way that might be expected if they were all true blood relations and, in the widest sense, members of the same lineage. Such are the arguments derived from the study of homology and taxonomy, from recapitulation, from geographical distribution, and from such particular aspects of heredity as hybridism. Then there are the evidences which are concerned with demonstrating that change or variation, of the kind without which evolution is almost unthinkable, does occur in plants and animals. Finally, there is the palaeontological evidence, or the fossil record of the rocks, which shows that throughout much of geological time there has been a grand succession of life-forms such as might be expected if an evolutionary progression had been at work.

Among these the last, palaeontology, occupies a special place, because it contains, in a way that the rest do not, the dimension of time, and thus goes far towards enabling us to comprehend, at any fleeting moment, the presumed results of evolutionary change over immensely long periods. It telescopes the events of successive past ages into the understanding of the present.

While the other lines of evidence indicate the likelihood of an evolutionary past with various degrees of probability, they reveal little or nothing which cannot be accounted for on some different hypothesis, but with palaeontology it is significantly otherwise. If the record of the rocks is reliable, and by and large it is believed to be so, then it is clear that different organic forms have, in the course of time, succeeded one another in a sequence of progressive change in a continuing direction, and this is as nearly as one can well approach to a simple statement of the leading criterion of evolution. At the same time it must be remembered that though the record tells us that the past has been a structurally orderly succession of types it does not, and cannot, tell us unequivocally whether or not this has arisen in the course of genealogy by the actual transformism of one type into the next which is evolution in the fullest and strictest sense of the term.

Unfortunately the palaeontological evidence is not equally satisfactory for all kinds of living things. The preservation of organic remains as recognisable fossils happens only under favourable and rather particular circumstances, in which bodies with robust skeletal designs are at a great advantage, and the fossil record is therefore largely the record of those groups which are for this reason, most copiously and perfectly preserved as fossils. Of other kinds of life some are less well preserved, and some, especially the softer plants and animals, are scarcely preserved at all. Indeed the palaeontological evidence for evolution is to a large extent derived from the fossils of vertebrates, which have bony endoskeletons, and from those of the shells of molluscs. Occasionally, it is true, peculiar methods of preservation, such as coal formation, tell us a good deal about other groups, including many early plants, but these cases are too infrequent and irregular to be of really comparable value.

Notwithstanding these limitations the fossil record is unquestionably the most satisfying of the evidences of evolution, not only because it is the most uncontrovertible, but because it is the most easily demonstrated to those who may not be accomplished biologists, and it was, from the beginning, one of the chief weapons of the evolutionists. Indeed, it may be looked upon as the origin of the whole more restricted biological theory since it was the impossibility of reconciling the fossil succession with current ideas of cosmogeny which more than anything else set men's minds thinking along new and original lines.

There are two outstanding historical causes of lack of balance in evolutionary studies, one being the general time-lag which has always existed between an understanding of parts of the animal kingdom and a corresponding understanding of the plant kingdom, and the other the more precise historical and chronological circumstances which attended the establishment of the Darwinian doctrine.

With regard to the first, man has never been in doubt that some at least of the other higher animals are alive in the same sense and in much the same way as he is, but the extension of the recognition of comparable vitality to other and simpler kinds of animals is more recent, because it presupposes a degree of anatomical and physiological knowledge which has only very gradually been acquired, and its further extension to include plants has been an even later step. Even to-day the phrase "natural history" often refers to animals only, and most books with this title are in fact devoted exclusively to zoology.

As for the second point, it is to be remembered that while the idea of evolution itself was no new thing, the large-scale conversion of thought to a belief in it began with the work of Darwin, and although it is now a truism that evolution and Darwinism are not synonymous, it may be doubted whether this distinction was appreciated in the earlier days, and whether it was fully realized that the directions of emphasis proper to Darwinism were not equally applicable to all aspects of evolution. It was rather a question of establishing the wider conception through that interpretation of it which is called Darwinism, and this, for three reasons, deeply affected the development of the whole subject.

First, the crucial exposition of Darwinism would probably never have achieved its rapid success had it not been couched in comparatively simple language, and had it not concentrated attention on many matters of common knowledge and observation about domesticated and other familiar animals. Second, the idea of *natural selection*,* which was fundamental to that exposition, could clearly be best represented, at any rate to an unaware public, by

* In this book the phrase *natural selection* is used in its original and proper Darwinian meaning as expressed by the author in the fourth chapter of *The Origin of Species* in the words "This preservation of favourable individual differences and variations, and the destruction of those which are injurious, I have called Natural Selection or the Survival of the Fittest".

referring especially to those organisms (the vertebrates) in which a relation between predator and victim is most obvious. Third, it was inevitable that the wider implications of the evolutionary thesis should sooner or later become focused on man and his nearer relations in the animal kingdom. Add to this the natural tendency to study those animals in which comparisons based on individual human knowledge and experience can most easily be made and it is easy to see why the study of evolution soon becomes the special province of certain branches of zoology.

But it cannot be supposed that these circumstances entirely account for the Darwinian achievement, for *The Origin of Species* is not so easy to read as to make it likely that it captured the public imagination of its day by itself. Clearly there were other and more subtle factors at work.

The remarkable success of Darwin's theory was undoubtedly due in chief to the fact that the world was ready for it. History has over and over again pointed the contrast between the fate of ideas born before the corporate human mind was ready for them and those which were fortunate enough to appear when what has been called *das psychologische Moment* was in their favour, and among these latter Darwinism is an outstanding example.

The century from 1750 to 1850 had seen, first in this country and later elsewhere, those profound social and economic changes which we now call the Industrial Revolution, and the time was in any case ripe for some kind of stock-taking, but towards the end of this period two closely successive phases went far to settle in advance the result this would have. The first was the great wave of industrial and political unrest which, abroad, culminated in the Year of Revolution (1848), and at home in the agitations of the Chartists, and which sprang, at least in the latter case, directly from the bad social conditions which the earlier fumbling days of industrialism had brought about. The second was the astonishing material dividend which, only a few years later, this same industrialism began to pay, a prosperity which was, though perhaps by contrast only, at its most obvious just about the time that Darwin's great book appeared.

These two phases caused a dilemma which must have been profoundly distasteful to many thoughtful people. The revolutionary movements had, almost for the first time, brought home to many who were but remotely concerned with them the terrible social conditions of the age, and did much to arouse a lethargic public opinion; but the later prosperity made it all the more difficult to see these things, and the necessity for remedying them, in the proper perspective. Could the basis of so much apparent good be so essentially bad, and if so, ought it to be condoned or ignored for the sake of its fruits? The problem was whether the ruthless competition which seemed inseparable from the existent economic system could be accepted with complacency or not, and we shall certainly do contemporary society a grave injustice if we do not believe that many of its members were acutely aware of it. Guidance, especially in the form of some apparently common-sense and rational resolution of the difficulty was urgently desired, and this Darwinism provided. According to this theory the struggle for existence and the survival of the fittest were to be seen, not uneasily as the ugly products of human cupidity, but with composure as an integral part of the great plan of Nature itself and thus, it might be assumed, as reflecting the sanctions of highest authority.

While there is no reason to suppose that Darwin's biological deductions and inferences were ever consciously coloured by the hues of contemporary thought it seems unlikely that he can have been totally unconscious of the trends of the time, particularly since he belonged to a section of society remarkable for its intellect, social awareness and industrial enterprise. It is to be remembered also that the idea of a struggle for existence was at least as old as Malthus, to whom Darwin readily acknowledges his indebtedness, and that Tennyson in his poem *In Memoriam* (LIV, LV) had re-directed public attention to it some years before *The Origin of Species* appeared (see Poulton, E. B., in *Charles Darwin and the Origin of Species*, London, 1909).

Of the more subjective influences which have affected the study of evolution only two need be referred to here, and that but briefly.

One is the question of what exactly the word evolution implies. Reference to larger dictionaries shows not only that it has many meanings and applications but that the one in which it is now most familiar (the biological) is among the most recent and derivative of them. Thus, there is every opportunity for the original and strict biological connotation of the word to become blurred by false analogy, and it is important in any book dealing with the subject to define the meaning which is held to attach to it therein. In this book the intended definition of the word *evolution* is that given in the *Oxford English Dictionary*, namely, "*the origination of plants and animals, as conceived by those who attribute it to a process of development from earlier forms, and not to a process of 'special creation'*".

The other, and more serious, objective difficulty appears to be psychological. For some reason which is not easy to understand the discussion of evolution and its many problems commonly evokes degrees of feeling unusual in scientific debate and more familiar in religious controversy. There seems an almost irresistible temptation to disregard the stricter canons of scientific argument, which, considering the meticulous care of Darwin himself in such matters, is as curious as it is regrettable. In consequence the literature of evolution has to be read with great caution, and even serious contributions to it are too often marred by a failure to let the facts speak for themselves, this failure usually taking the form either of a tendenciousness of phrase or, more frankly, of emotive appeal. This is why so much of the literature is less convincing than it might be and why there is so often produced in the minds of its readers that confusion and sense of unreality which has already been mentioned.

As a result of all these factors our present-day picture of evolution is in many respects imperfect because it is painted too largely in the colours of certain kinds of living things only. This unevenness, which is an expression of relative lack of knowledge, is seen on the large scale in the general disparity between the study of evolution in the plant and animal kingdoms. It is true that the subject of palaeobotany has been highly developed and that most general accounts of organic evolution briefly describe the fossil history of plants, as well as the succession of increasingly differentiated types in time, but it is seldom that any existing plant group has been studied primarily from the point of view of the evolutionary processes which may have produced it, and not merely as the concomitant of the establishment of some natural classification. Indeed studies in plant evolution are generally no more than the raw materials of phylogenetic speculation, a mental exercise which is valuable in moderation but which is easily carried to excess. Certainly no considerable part of the plant world has been studied by those interested in wider evolutionary problems, with the same breadth and depth as have some groups of the animal kingdom. This is not to maintain that plants have never been the subject of such investigations for the botanical work of Darwin himself is enough to disprove this, but this very point is in fact significant because this was mainly the product of his maturer years, and marked a turning away from animals towards the plants which, he makes clear, attracted him because of their comparative simplicity (see Darwin, F., ed., *Life and Letters of Charles Darwin*, London, 1887, vol. III, pp. 244 and 256).

On a smaller scale there is noticeable lack of balance *among* the groups of the animal kingdom, a quite disproportionate emphasis being placed on the vertebrates. This was no doubt justifiable, and probably unavoidable, in the early days, but its perpetuation has been detrimental for several reasons. These are the animals in which brain development is highest, and they include the most intelligent of all, man himself. Inevitably there is a tendency to attribute to them, and, by extrapolation, to the whole theory of evolution, many of the impulses and influences of human experience, so that the language of the subject has gradually acquired an idiom of human behaviourism for which there is little justification. As a result evolution, as it is commonly pictured today, is largely evolution in terms of those animals which are most sentient; which are endowed with most volition and desire; and which are most capable of satisfying these. It is concerned to an inappropriate degree with those living things in which the innate evolutionary machinery is most obscured by the development of mentality, which, in the broadest view is something superposed upon the fundamental pattern of evolution. It is, however, still the recognition of this fundamental pattern which is our primary aim and so it may well be that the study of those living things which are less well-endowed in this particular respect will prove more profitable.*

Preoccupation with vertebrates has also been detrimental

---

* In this connection it is not altogether irrelevant to remind readers that anthropology, which is the study of the evolution and history of the human species, is notoriously contentious.

because it has caused exaggerated stress to be laid upon the "predator-victim" relation, and, through this, on the conception behind such phrases as the *struggle for existence* and *survival of the fittest*, so that these have become, too often and too widely, to be regarded as essential expressions of evolution, an error which has had a profound effect on human affairs far beyond the limits of scientific biology.

Lastly, this concentration on animals has almost inevitably led to the implications that plants are, presumably by their very nature, of less consequence and therefore less valuable in the study of evolution than animals, and this has in its turn served to intensify the very concentration through which it arose. We shall see in a moment that there are good reasons why this implication is not as strong as it used to be, but it is still all too common to find an almost complete neglect of plants in what are presumably intended to be general writings on evolution.*

This lack of balance in the study of evolution was probably at its greatest towards the close of the last century, because since then the growth of the great science of genetics has brought about a considerable measure of readjustment. One reason for this is that the new demonstration afforded, by genetics, of that mutability which is the raw material of evolution is most easily to be observed in organisms which are suited to experimental breeding, and among these the shorter-lived flowering plants are very important. Another reason is that genetics has shown the widespread existence in plants of the hitherto unsuspected phenomena of polyploidy and aneuploidy (see next chapter).

However, invaluable as genetics has been in this way, it has its own quite natural limitations, and that these are not always fully discounted is shown by the common tendency to use the word *evolution* in contexts which suggest that it is synonymous with *genetics*. It must also be remembered that genetics, while revealing much about the fate, in the course of inheritance and otherwise, of characters already in existence, reveals less about what is really the core of the evolutionary problem, namely *in what circumstances*

---

* As for instance in G. G. Simpson's *The Major Features of Evolution*, New York, 1953. Despite its title this book of 400 pages contains only about thirty direct references to plants, most of them very brief and many of them quoted from a single source (G. L. Stebbins, jr., *Variation and Evolution in Plants*, London, 1950). It dismisses the outstanding phenomenon of polyploidy in a few lines.

and for what reasons novel characteristics make their appearance in nature. To this extent, at least, genetics is the study of effects rather than causes, and so, though it can be very valuable in this way and though it has done much to redress the balance of interest in evolution, it has not affected the position as a whole as much as is sometimes supposed.

It has been argued that the position today with regard to the study of evolution is not as satisfactory as might be expected because of a lack of balance between the different aspects of the subject, in consequence of which the fund of factual knowledge on which opinions about evolution must necessarily be based is far from complete. While much is claimed to be known about evolution in some kinds of living things, and notably in certain animal groups, much less is known about it in others, and this is especially true of much of the plant kingdom.

If this diagnosis of the situation is correct then it is plain that the remedy is to extend investigation more widely into those groups of the living world which have so far received inadequate attention, and this book is a modest attempt to do something of this sort in respect of the great group of the Flowering Plants. The method adopted is to apply as widely as possible the principle that one of the best ways of learning more about causes is to study their effects with greater care. The impossibility of determining evolutionary causes by direct observation has already been stressed and evolution is therefore not only a subject in which there is special scope for the application of this principle, but one in which it promises to be particularly useful. If we are to gain any real understanding of the *course* of evolution we must first appreciate to the full its *achievement*, for the more clearly this is realized the more hope there is of finding out how it has been brought about, and it is from this point of view especially that the Flowering Plants are approached in the following chapters.

That the Flowering Plants or Angiosperms* have not received more attention from those interested in evolutionary problems is doubtless in the main due to the enormous and forbidding size of the group but it is partly also due to the fact that though these

---

* In this book the term *Flowering Plants* is generally used in preference to, but as synonymous with, the more scientific words *Angiosperms* and *Angiospermae*.

plants have a considerable fossil record this is not of the kind which gives much direct information about phylogeny. In fact neither of these circumstances is as discouraging as might be thought. Even though, perhaps, no one can hope to encompass such a vast array of life in extreme detail there is much to be learnt from more general surveys, and provided these are comprehensive enough there is nothing unscientific about this kind of treatment. On the contrary rather, had some of the earlier workers in evolution adopted a broader approach to their subject our difficulties today, even if no less teasing, might well be fewer. Moreover the very size of the group gives opportunity not only for such broader casts, but for many lines of more intensive investigation, such as might emerge from an attack on a wider front. The history of the Flowering Plants is, true, comparatively short, but this in combination with their great numbers suggests that they show, not a residue of evolution illustrated by but a few disparate relic types, but evolution on the grand scale and as a going concern, and quite possibly in one of its most active expressions. There is also the advantage that its earlier chapters, though at present obscure, are not irrevocably lost in the mists of antiquity.

In many other ways too the Flowering Plants provide promising material for evolutionary study. They comprise large numbers of systematic units or *taxa* of all values, and thus give ample opportunity for any quantitative treatment that may be desired. The higher categories among these taxa are not so unlike one another as to raise legitimate doubts as to their true consanguinity and yet, within this reservation, they include examples of all sorts and of all grades, so that more qualitative appreciations are also easily possible. The group is copiously and very evenly represented in all kinds of climate and habitat, so that due account may be taken of environmental factors and influences: and its members show the nutritional methods normal for their kingdom. Finally speciation in them is, to a very high degree, based on characters which are generally independent of physiological condition, that is to say on characters which are not plastic, and this, as will be seen in due course, is a matter of great importance.

CHAPTER TWO

PLANTS AND ANIMALS

IT IS implicit in the argument developed in the last chapter that the lives of plants and of animals are sufficiently akin to warrant the assumption that they are but different expressions of the same phenomenon. At the same time it is a matter of common observation that there are profound differences between the two kingdoms. This combination of similarity and dissimilarity is clearly of great importance to those who study evolution because their whole approach to that subject may be influenced by their assessment of it, and for this reason a somewhat full presentation of the situation is made the topic of this chapter.

In making it, our attention may safely be confined to the "highest" members of each kingdom only, namely the Flowering Plants on the one hand and the Vertebrates and Insects on the other, because almost everything that can be said of types lower in the presumed evolutionary scale can be said *a fortiori* of these groups, while it also avoids any sterile discussion about the plant or animal affinities of simpler organisms. Unless there is some particular qualification then, the word *plants* here means the Flowering Plants and the word *animals* refers to the terrestrial vertebrates and insects. At the same time it will be sufficiently apparent that, in fact, most of the statements have a wider application. It is also convenient, in view of the rather complicated inter-relationship between many of the points to be mentioned later, to make a brief and formal statement before going on to a more detailed consideration.

A Summary of the Similarities and Differences between Plants and Animals

1. Plants and animals are like one another, and unlike anything else, in possessing protoplasm, and this is why they are basically similar in respiration, assimilation and irritability: cellular construction; sexual reproduction; and in taxonomic constitution.

2. Plants and animals are profoundly unlike one another in their modes of nutrition (functional) and of growth (formal), plants being immobile-passive and animals being mobile-active.

3. To this fundamental distinction of behaviour may be ascribed most, if not all, the other differences between plants and animals, which, with one exception, are comprehended under the headings:

Biological—Numbers of offspring; dissemination; dormancy; alternation of generations, and time of meiosis.

Cytological—Osmosis, cell walls and vacuoles; intercellular substances and solidity of tissue; ergastic substances and protoplasmic inclusions; centrosomes and cytokinesis.

Reproductive—Sex distribution and sex chromosomes; male gametes, double fertilization, polyandric progeny and hybrids; polyembryony and related aberrances; cytoplasmic inheritance, chimaeras and graft-hybrids, and vegetative propagation.

4. Polyploidy (including aneuploidy). The common occurrence of polyploidy in plants but not in animals, is a phenomenon of such great interest and significance that it calls for particular emphasis. This, also, may be attributed to more fundamental differences, though it seems one of the most isolated distinctions between the two kingdoms.

The Protoplasmic Similarity and its Consequences

That in both plants and animals life resides in the substance or chemical system called protoplasm and nowhere else is the most fundamental and familiar, but one of the least obvious, of the similarities between them. Protoplasm, though perhaps almost infinitely variable in chemical make-up, is essentially one in all living things; it is seen only in living bodies; it is the physical basis for all their vital processes. It is true that some of the properties of protoplasm may be more stressed in plants than in animals or *vice versa*, and there are doubtless many finer points of chemistry in which the two may differ, but so unique is protoplasm in itself, and so completely apart from any other chemical system, that such discrepancies amount to little when weighed against the underlying unity of organization deriving from its occurrence in both kingdoms.

Biochemistry: respiration, assimilation and irritability

Because the vital processes in both plants and animals are based on this substance protoplasm, there is a general similarity in their biochemistry, both showing the same physiological processes of respiration, assimilation and irritability. Life, in this sense, is the same in both, and in both the body machinery works in the same way.

This general point may be illustrated by one particular example, the relation between *chlorophyll* in plants and *haem* in animals. Chlorophyll occurs in the form of two closely related and associated substances Chlorophyll a and Chlorophyll b, which are the chief contributors to the green colour in plants. These have complex molecules, each containing many carbon and hydrogen atoms, a few oxygen and nitrogen atoms, and one atom of magnesium. Haem is the pigment of the red blood corpuscles in animals and plays a physiological part in their lives paralleled only by that of chlorophyll in plants. It, also, has a complex molecule, containing many atoms of carbon and hydrogen and a few of oxygen and nitrogen, but in place of the magnesium found in chlorophyll it has an atom of iron. Furthermore, chlorophyll and haem have, as a common basis, a porphyrin structure.

Cellular construction

That both plants and animals are built up of units called cells and that these are arranged in tissues, each consisting largely or entirely of one particular sort of cell, is so familiar that it is easy to forget what a peculiar circumstance it is. Since it is universal it is hard to imagine any other state of affairs, but it is well to remember how complex and characteristic this kind of organization is. That two series of living things should exhibit, virtually in complete parallel, such a condition may not unfairly be held to amount to almost the strongest possible circumstantial evidence of their community of origin and of their real relationship. Association of the cellular condition with protoplasm can scarcely be regarded as other than that of effect and cause, in which the cell is to be looked upon as the framework within which protoplasm functions, and its occurrence in both organic kingdoms is a strong intimation that the substance behind it, whatever variation in detail it may show, is fundamentally the same throughout. Moreover, the size of both plants and animals is increased by a multiplication of their cells, and the division of the protoplasts which precedes this cell multiplication is carried out in both by the same highly particular and characteristic process of mitosis, which involves the nice duplication and redistribution of the chromosomes. On the other hand the differences of detail in the cellular organization of plants and animals take the form of modifications of a common plan in accordance with the distinctive overall modes of life of the two: in no case are these differences really fundamental in kind.

Sexual reproduction

The similarity between plants and animals here has three main aspects—the overall likeness of nuclear process and structure, including the broader features of chromosome morphology and its relation to inheritance; the general identity of the principle of meiosis; and the general parallel in the form of the gametes, in the process of their union, and in their occasional elimination.

Taxonomic constitution

Taxonomic classification is identical, both in principle and in application, in both plants and animals, and this general resemblance, which runs throughout the two kingdoms, is also perhaps to be looked upon as deriving from the underlying unity of cellular organization, though clearly in a more subjective way, since formal taxonomy is man-made. It has a more artificial expression in the successful use of the same hierarchy of categories in both animals and plants, and a much more natural one in the intriguing parallel which can be detected at many points between the members of the two kingdoms.

For example the archegoniate and seed plants may be thought of as the botanical counterparts of the terrestrial vertebrates or arthropods; the seed plants alone may be thought of as corresponding to the higher vertebrates; while the algae are clearly paralleled by the coelenterates and some other simple marine groups. Within the higher groups too there are striking parallels, as between the archegoniate cryptogams and the amphibia and reptiles; between the gymnosperms and the birds and oviparous mammals; and between the angiosperms (flowering plants) and the placental mammals.

This in turn leads to another most interesting comparison, namely that in both plants and animals one of the most conspicuous themes of structural development has been, to judge from the succession of life revealed by the fossil record, the increasing protection of the sex cells and young embryos against the difficulties inherent in a subaerial (atmospheric) environment, and the elaboration of an appropriate soma in correlation with the physiological changes involved in the exploitation of an environment of this sort. In short in both plants and animals the occupation of the land, which seems to have been the main theme of their evolutionary story, has followed much the same course and has been achieved by what are fundamentally much the same methods.

The Physiological Differences between Plants and Animals

Functional

The most obvious difference between plants and animals is the immobility and passivity of the former as compared with the latter, a difference which is of great importance because most other distinctions relate to it. This difference of behaviour reflects the fact that while green plants can manufacture the food they need from certain raw materials in the medium of air and soil in which they find themselves, animals can make no corresponding direct use of their surroundings and must obtain their food in some already elaborated condition from places where the previous activities of plants have made it available. It may be said that "food comes to the plant" but "the animal must go to its food", and as a result of this the lives of animals are to a large measure engrossed in activities unknown to plants, the search for and pursuit of food supplies.

The effect of this situation on the grand design of nature has been decisive. In consequence of it the whole system of life on land rests solidly on that great section of the plant kingdom which is dominant today, and with which we are concerned particularly here, the Flowering Plants. With few exceptions, everywhere that life exists in quantity, except in salt water, these plants are its foundation; they provide most of the organic constituents of the soil on which so much saprophytic life depends; they directly nourish the herbivores; and these in turn become the food of more predatory kinds of animals. Only in one major section of world vegetation, that of the coniferous forests, does the plant life as a whole belong predominantly to another group, the Gymnosperms, and even here the alien dominance, though conspicuous, is far from total. Elsewhere only in such purely local situations as may be illustrated by a bracken-covered hillside, or in conditions so stringent that nothing more demanding than simpler plants can survive, is there to be found any plant cover (and the word is deliberately chosen) which is not almost wholly composed of flowering plants. These plants must therefore be recognized as constituting the foundation of life on land, and in virtue of their consequent standing in the general scheme of nature, as influencing, in some ways at least, all other kinds of living things.

This physiological difference not unnaturally bears on many more subjective problems, of which one may serve as an example. The pursuit of food by animals presumably entails the exercise of choice, since, it may be expected, there will be at any time various courses of action open in this respect which, though possessing no inherent order, can only be implemented in succession. Similarly it may be said with truth that, in comparison with animals, plants are non-sentient, being as far as we know without anything corresponding to the nervous system of animals or to their instincts, intelligence, and volitional satisfaction of desires, at least in so far as this last may be defined as an urge or instinct to select one presented course of action rather than another.

Formal

Although the difference between passivity and activity was rightly described as one of the most obvious, this might even more truly be said of the distinction in form which accompanies it, because the whole structure and organization of the body is distinctive in a way which is not always fully appreciated. While in animals growth takes the form of a gradual expansion of a single unity into a definitely proportioned body having a strictly limited range of variation in dimensions and ratios, growth in plants is essentially and indefinitely accumulative, taking the form of the multiplication and apposition of basically similar units of construction. In consequence the individuals of an animal species vary little in size and even less in shape, but the individuals of a plant species may vary enormously in size and considerably in shape. Indeed it is no uncommon thing to find within a single species of flowering plant individuals so small that they consist of only one organ of each sort and individuals in which the number of these stems, leaves and flowers, may be hundreds or even thousands. Nor is there any inherent limit and plants tend to continue to accumulate organs so long as they live and grow.

The all-important point about this is that the plant method is an indefinite one, and that the size of a plant depends, for this reason, and in a way which is not true of animals, on the number of increments it can grow, which is normally a function of the time and supplies available. This means that a relative deprivation of supplies which would cause lethal results in animals may, in plants, cause no more than a marked variation in size. The effect of this on the economy of nature is manifest. Plants, being immobile, are unable to move from points of scarcity to points of abundance as occasion demands, and the fullness of their lives is rigidly ordained by the accident of their situations, and the nutritional flexibility provided in animals by their ability to search out and thus augment their food supplies is replaced in plants by a principle through which size may be adjusted to the measure of the resources which happen to fall within the individual's command.

The More Consequential Differences between Plants and Animals

The many secondary differences between plants and animals may largely, and perhaps entirely, be regarded as arising directly or indirectly from the fundamental distinctions of function and form which have been discussed. They may be grouped under four heads, of which three can be described as biological, cytological and reproductive, and of which the fourth is polyploidy.

Biological

These appear most plainly as differences in life-cycle and they have four chief aspects.

1. Number of offspring

Owing to the peculiar structural organization of plants individuals may vary greatly in the number of flowers they bear, but generally speaking the flowers of a small individual, though normally fewer, are not appreciably smaller in themselves. This we may suppose reflects their reproductive function and the fact that cells do not vary greatly in size, but the result is that the number of progeny of any plant depends almost directly on the number of its flowers and may therefore be enormously variable, certainly much more so than is usual in animals. The number of flowers borne on an individual is, moreover, a measure of nutritional resources and hence the progeny of plants are much more closely and directly related numerically to the availability of resources than they are in animals. In reproduction plants thus show the same greater flexibility that was noted in the matter of size.

2. Dissemination

It is an interesting distinction between plants and animals that in the latter it is normally the parents which spread the stock by movement and the newly born young which are most sedentary. In plants on the contrary the parents remain rooted in the soil (except in the comparatively rare tumble-weeds) and it is the very young offspring which are dispersed. Dissemination may therefore be said to occur before or as a preparation for parturition in animals, and as a sequel to parturition in plants.

3. Dormancy

In most plants the embryo, at a comparatively early stage in its life, passes into a dormant state which may continue for years, a state of affairs never exactly paralleled in animals. This dormancy provides a period during which the seed or fruit may become dispersed, and is a most interesting instance of the essential passivity of plants. It is difficult to imagine a similar condition arising in warm-blooded animals, and even though something of the kind occurs in hibernating reptiles and arthropods and in some molluscs it is scarcely directly associated with dissemination, except perhaps by accident. In plants, it would seem, time for chance and opportunity takes the place of directive activity.

4. Alternation of generations and time of meiosis

In both plants and animals there is an alternation of haploid and diploid phases, but there is a basic distinction between the two cases. In animals the diploid phase ends during gametogenesis and the succeeding haploid phase lasts only until syngamy, so that it is solely concerned with the elaboration of gametes. In plants the diploid phase ends with the formation of spores, and the succeeding haploid condition lasts from the production of spores, *through* the gametophytic tissue and gametogenesis until syngamy, so that there is a diploid soma concerned with the production of spores which will, in their due course, give rise to gametes. In flowering plants the "plant" is therefore a sporophyte and the gametophytic generation is extremely short, but in animals there is nothing corresponding to the sporophyte at all. The usual explanation of this is that in the course of the development of plant life on land the need for greater protection in the new environment led to the retention, on the body of the parent, of the female gamete which had hitherto also been the agent of dispersal, and that this loss of motility was in due course made good by the elaboration, for the purpose of multiplication and dissemination, of that diploid phase of the life-cycle which is now called the sporophyte or sporophytic generation.

Cytologically this difference is, essentially, that of the time at which meiosis takes place. In plants the meiocytes are the sporocytes but in animals the meiocytes are the primary gametocytes

Cytological

1. Osmosis, cell walls and vacuoles

The essence of the physiological differences between plants and animals is that the cells of the former are much more highly osmotic than those of the latter, and, indeed, it is common to treat osmosis as a botanical phenomenon, though it would be wrong to suppose that there is nothing of the sort in zoology. At all events the osmosis of plant cells is of a much higher order, and from this circumstance comes the distinction so often canvassed as a final criterion between plants and animals, that the former have rigid cell-walls of carbohydrate material and the latter non-rigid membranes of a protoplasmic, or at least protein, nature. The presence of a rigid cell-wall limits the osmotic expansion of the protoplast and this in turn sets up high pressures within it which express themselves as turgidity. This turgidity is the counterpart in many land plants of the rigidity which is provided in animals by the muscular and skeletal systems, and is thus another aspect of the passivity of plants.

Associated with the more active osmosis of plant cells is the presence in them of large static vacuoles, which are often their most characteristic feature and which not infrequently result in the protoplasts being little more than bags of liquid. Animal cells on the contrary have no comparable continuous vacuoles.

2. Intercellular substances and solidity of tissue

In large organisms, such as many higher plants and higher animals, the limiting walls or membranes of some of the cells usually become, in time, mechanically insufficient and require to be reinforced, and the way in which this is done normally reflects the osmotic differences between the two kingdoms. In plants the cell walls become thickened on the inside in a way which allows the cells to continue to behave osmotically as long as they remain alive, and adjacent cells remain in their original close contact, but in animals the common circumstance is that material is secreted *through* the cell membranes and accumulates, often in almost solid form, *between* the cells, thus thrusting them more widely apart and forming a matrix through which they become spaced.

It is these features, both of which are associated with osmosis, which most contribute to the characteristic distinctions of texture and solidity in the living tissues or "flesh" of plants and animals. It may also be noted that this difference of physical consistency explains, in so far as it relates to osmosis, the rather unexpected fact that plant tissues are often more difficult to grow in isolated pure culture than animal tissues.

3. Ergastic substances and protoplasmic inclusions

Because of the fundamental physiological difference it is not surprising that the organic substances found in the cells of plants and animals respectively as a result of their metabolic activities, or ergastic substances as they are called, are often very different, and would no doubt appear to be even more so were it not that some of the substances identifiable in animal cells have come there by absorption from plant tissues. As the briefest possible more factual statement it may be said that carbohydrates, and especially sugars and polysaccharides, are particularly characteristic of plant cells, while they are much less conspicuous and tend to be of rather different kinds in animal cells, starch being little in evidence. This state of affairs is often related to the more active osmosis in plants and to the circumstance that the sugars and starches are respectively osmotic and non-osmotic carbohydrates, so that conversion from one to the other offers a ready means of controlling osmotic values.

Less closely associated with osmosis are the differences found among the inclusions of the protoplast in the two kingdoms. In animal cells there are various structures, which generally go by the name of "Golgi apparatus", which are not to be seen in plants, but the most remarkable inclusion is one found in plants and not in animals, at least in comparable form, namely the plastid. These are bodies in the protoplasm which are associated with the presence of chlorophyll and starch formation. Chloroplasts, which are those plastids containing the chlorophyll, are the best known and most easily studied kind and they may almost be considered as the very seat of the fundamental physiological difference between plants and animals.

4. Centrosomes and cytokinesis

The general similarity of mitosis, or nuclear division, in plants and animals has already been commented upon on p. 16, but associated with this process are some minor differences between the two. For instance some of the finer points of chromosome rearrangement may not be identical, and telokinetic chromosomes are said to be rare in plants, but there are two rather more considerable points.

In higher plants the figure in nuclear division, the mitotic figure, is anastral, but in higher animals it is amphiastral, this being associated with the presence in the latter of a centrosome which divides before the rest of the nucleus and passes to the ends of the nuclear spindle.

The other point is in cell division proper (cytokinesis) which follows mitosis. Typically in plants the new dividing wall between the daughter protoplasts is laid down in its entirety on a middle plane of the parent cell soon after the daughter nuclei have diverged, but, as typically, in animals, cytokinesis is by the inward growth of a furrow from the periphery of the protoplast, this gradually constricting the latter into two.

Reproductive

1. Sex distribution and sex chromosomes

The hermaphrodite condition is much more common in plants than in animals and *vice versa*. The proportion of flowering plants in which there is strict dioecism, that is to say the occurrence of unisexual individuals, is difficult to compute because some appear to be functionally though not structurally so, but in terms of families it may be said that only about 20 out of upwards of 400 are entirely dioecious; that only in about 40 others does dioecism occur at all; and that nearly all these 60 families are relatively small. On the other hand in the higher animals the unisexual condition is almost or quite invariable. Not only this but in plants various other conditions are to be seen in addition to these two. Monoecism (separate male and female flowers on the same plant) is a good deal more prevalent than dioecism, while at least in thirty families there occur some or all of the various sexual combinations between male, female and hermaphrodite that go by the name of polygamy. In short, while in animals sex distribution is constant, it is far from being so in plants.

The nature of the sex chromosomes is also not precisely the same in both plants and animals. Each exemplify both the XY and XO types, but in plants there are certain others as well, including the YXY, and this state of affairs is perhaps correlated with the much greater rarity of the unisexual condition in plants.

2. Male gametes, double fertilization, polyandric progeny and hybrids

In the actual sexual process, which may be taken to comprehend the gametes and their fusion, the picture is not very easy to draw because there are a good many variants in gametogenesis in both kingdoms, but for the present brief consideration the essential points are three. In higher animals the male gametes are highly complex and very active sperms, but in flowering plants (and most gymnosperms) the male gametes are simple non-motile nuclei which are almost passive and which may or may not be immersed in some surrounding cytoplasm. Similarly the actual details of syngamy are diverse, but here there stands out that most remarkable phenomenon of double fertilization, which occurs in the Flowering Plants only, and one result of which is the initiation of their unique triploid endosperm tissue.

The third point is the absence of active copulation in plants and this may greatly have influenced their historical development. In flowering plants the stigmas are collecting surfaces, and as regards the actual process of this collection make little or no selection among the various pollen grains which may present themselves, except perhaps in so far that some designs of pollen grain may not be able to adhere to particular types of stigmatic surfaces. It is true that various internal mechanisms may condemn certain pollen, which has adhered, to subsequent nullity, but even allowing for this it is clear that the procedure in plants makes possible a far greater degree of variability within the total progeny of any one generation (which may be expressed more simply by saying " among the seeds of any one fruit ") than in animals. In copulating animals the members of any single litter or brood are, normally at all events, full brothers and sisters, but in plants it may be more usual than otherwise for this not to be the case. Not only this but there is little which can be held to guarantee the exclusion of what may be called "foreign" pollen from functioning in fertilization, so that there may be, among the progeny of a parent plant at any single reproductive phase, not only full brothers but half-brothers also, and the latter may include various interbreeds and hybrids. Indeed the often-stressed greater frequency of hybrids in plants, with all its evolutionary implications, is presumably partly due to this, though it is also strongly associated, as we shall see, with polyploidy.

3. Polyembryony and related aberrances

The position here can be described in the simplest way by saying that, while the particular form of apogamy known as partheno-genesis, which is the development of a female gamete without syngamy, occurs in some of the arthropods as well as in plants, there are two, more extreme, forms of apogamy which are found only in plants. These are polyembryony, which is the development of more than one embryo in the embryo-sac, and adventitious embryony or sporophytic budding, which is the development of an embryo from a cell or cells of the nucellus or even of the integuments. Both these latter structures belong to the sporophyte generation and hence to a phase not represented in higher animals.

4. Cytoplasmic inheritance; chimaeras and graft hybrids, and vegetative propagation

The presence of coloured plastids, especially chloroplasts, in the cytoplasm, often expresses itself in the superficial tissues of plants by easily visible characters, such as the distribution of colour, and some of the facts relating to these characters have suggested that there is, or may be, in plants, a special kind of inheritance mechanism which does not so directly involve the nucleus as usual, and which may be called cytoplasmic inheritance. The facts about this are as yet incompletely known but if some interpretations of them are correct it may be that in plants there is some kind of machinery by which characters can be transmitted from one generation to another which is not represented in animals, at any rate in quite the same manner or degree, and if this is true its implications may be very far-reaching.

In so far as chimaeras may be the results of polyploidy or of graft-hybridization, they are presumably confined to plants, and it is unnecessary to debate whether the second of these is ever an entirely natural process. Vegetative propagation is a simpler issue and the remarkable readiness with which many species of flowering plants increase the number of their separate entities by some method of this sort, is not only a normal and natural procedure, but one, again, unparalleled among the higher animals.

Polyploidy

Polyploidy may be defined as the multiplication of all or part of the genome, and since its evolutionary significance is obvious, it is not surprising that it has received a great deal of attention. It occurs very commonly in higher plants (especially in the Flowering Plants) but it is almost non-existent in higher animals, and it need only be said, and noted with care, that it endows the plant lineages which possess it with an entirely additional method of speciation which animals lack. When it is said that according to some calculations three out of every four species of Flowering Plants are polyploids of one sort or another, using that word in a wide sense, its significance can well be realized. It may be added that certain circumstances of polyploidy suggest that it much facilitates hybridism, and that this may indeed account for the greater numbers of hybrids among plants.

A particularly notable point about polyploidy is that it seems less directly attributable to the more fundamental distinctions of metabolism and form between plants and animals than most other secondary differences, and it is not easy to understand why it should be so characteristic of plants and so rare in animals. There are several ways in which it may be caused, the most straightforward, and probably the most frequent, being the failure of meiosis. This suggests that polyploidy may have some connection with the meiotic phase and perhaps with the difference of time of meiosis in the two kingdoms. If, as one school of thought believes, the sporophyte generation in higher plants has become intercalated into the life-history then it will have had the effect of separating meiosis in time from gametogenesis, with which it was formerly directly associated, and also of causing it to result, directly, in spores and not in gametes. Whether the twin influences of these changes of stage and product have, in the past, given the Flowering Plants in particular (in which the gametophytic generation is in its most reduced state) an inherent tendency or liability towards the suppression of meiosis, has not been demonstrated, but it is at least an interesting possibility.

Conclusion

The foregoing review shows that whatever the differences between the two kingdoms, these are accompanied by very striking similarities, so that the balance is clearly not all to one side. There is one series of what may properly be called fundamental resemblances; there is one profound sort of difference; and there is a whole series of what, in proportion, can only be described as lesser points of difference. All these items may, within the strict terms of our enquiry, which concerns only the highest types of plants and animals, be regarded as absolute, that is to say without noteworthy exception, and it is thus a matter of deciding their relative values.

This can best be done by widening, for a moment, the scope of comparison to include not one particular part of each kingdom only, but the whole of both of them. When this is done it is at once apparent that the resemblances remain unaffected and absolute but that the differences lose much of their sharpness. Thus, there are no plants or animals, in the usually accepted sense of these words, which have not a basically similar biochemistry; or which do not respire and assimilate; or which are not of cellular organization; or in which sexual reproduction is replaced by something entirely novel. On the other hand there are many forms of life commonly assigned to the kingdom of plants, which have not the characteristic passivity of plants, and many forms assigned to the kingdom of animals which do not show the activity typical of that kingdom, and it is not without deep significance that, usually in these cases, the accompanying distinction of growth form which might be expected is also much less clear, as may be exemplified by the motile unicellular algae and the bryozoa. Similarly with the lesser points of difference, which are either directly and invariably relatable to the more basic difference of passivity-activity or are found to be much less absolute. For instance, osmosis is seen in fullest expression only in the tissues of plants; polyploidy and centrosomes, on the other hand, characteristic though they are of one kingdom, do occur either occasionally or more frequently, in the other. In brief, it would seem that so long as features or items of this sort are not too inevitably the expression or consequence of other characters peculiar to one kingdom, they may be expected to appear, somewhere, among both plants *and* animals.

Finally there is the fact that, so long as the living world is regarded as comprised of two kingdoms only, plant and animal, certain groups of biota can find a place in one part of the scheme rather than another only by the exercise of arbitrary definition. Some of these organisms possess what are generally considered to be plant and animal characters respectively in equal measure, and can therefore be classified only by a decision to regard one or more of these characters as diagnostic; others show a level of differentiation so low that it cannot adequately be expressed in the language either of botany or zoology. In short, the differences between plants and animals are, in general terms, a function of their degrees of complexity and differentiation, which means that the higher plants and higher animals are more distinct from one another than are the members of any other parts of the two kingdoms.

All this seems to admit of but one conclusion, namely that of all the items of comparison and contrast which have been set out, one is immeasurably of more consequence than all the rest, this being that the basis of life throughout the organic world is protoplasm. There is clearly, here, a unity of life between the two kingdoms which outweighs any and all the discrepancies. From this it follows that the difference between the two in mode of life and structure, though in some ways the most remarkable item in the list, is no more than a difference of expression. Plants and animals exhibit two distinct economies of life but they do not exhibit two kinds of life itself. Where there is similarity between these economies there are associated likenesses; where there is little or no similarity there are associated differences. Plants express their liveliness passively and autotrophically, while animals express it actively and heterotrophically, and the relation between the two is part of the fundamental plan of nature.

Furthermore, this familiar physiological relation between plants and animals, by which, because of their inability to synthesize carbohydrates, animals are ultimately dependent for their existence on a sufficiency of plant life, makes almost unavoidable the conclusion that if, as is the belief most commonly held, life originated on the earth once only, its primaeval expression must either have been some one simplified condition from which both later kinds of organisms sprang, or must have had much in common with plant life as we know it in some of its simplest manifestations today. In either case it would seem that the two kingdoms must both have come, directly or indirectly, from one and the same source, and, consequently, that their later differentiation has been, in one sense at least, no more than a superficial divergence.

All this being so it must surely further be concluded that evolution, in the sense that change is a primary function of life, shows a comparable community between the two and that such differences as may exist in this respect are, like so many others, no more than consequent upon the distinctive patterns of life in the two kingdoms. There may be, and doubtless are, facets of organic evolution which are not found in both plants and animals but on the argument of this chapter these are not likely to be the more broadly revealing ones. It may, indeed, well prove that plants in general, and certain kinds of them in particular, display some at least of the features of evolution more plainly than do animals, just because they are, by comparison, non-sentient and passive, and because, in them, the expression of evolution is likely to be less overlaid by obscuring secondary effects than it is in animals.

NOTE ON LITERATURE

There is a useful brief survey of the subject matter of this chapter by Turrill, W. B., in "Discussion on the systematics of plants and animals and their dependence on differences in structure, function and behaviour in the two groups," *Proc. Linn. Soc. London*, 153rd Session, 1940–41, part 3.

On some of the more particular aspects of the matter the following, among others, may be consulted:

Alexander, J.: *Life: its nature and origin*. New York, 1948.

Jones, W. N.: "Chimaeras: a summary and some special aspects." *Botanical Review*, 3, 1937.

Lorz, A. P.: "Supernumerary chromosome reproductions: . . ." *Botanical Review*, 13, 1947.

Seifriz, W.: *Protoplasm*. New York and London, 1936.

Sharp, L. W.: *Fundamentals of Cytology*. New York and London, 1943.

Stebbins, G. L., jr.: "Types of polyploids; their classification and significance." *Advances in Genetics*, 1, 1947.

Vandel, A.: "Chromosome number, polyploidy, and sex in the animal kingdom." *Proc. Zool. Soc. London*, Ser. A, 107, 1937.

White, M. J. D.: *Animal Cytology and Evolution*. Cambridge, 1945.

CHAPTER THREE

MONOCOTYLEDONS AND DICOTYLEDONS

THERE are, broadly speaking, two sorts of flowering plants, which can be recognized, almost without exception, by the presence or absence of certain characteristics. This state of affairs is represented in taxonomic form by giving to these two sorts names based on one of their most notable differential features, the number of seed leaves in the embryo, and thus dividing the whole phylum into the two familiar series, Monocotyledons and Dicotyledons. That this segregation of the Flowering Plants into two can be made with so little difficulty is one of the most remarkable points about them, but like most other taxonomic dichotomies, it not only expresses the facts too formally but also gives an impression of fundamental importance which it may or may not deserve. Because of it the common conception of the Flowering Plants is that of an entity composed of two profoundly and cleanly distinctive, though unequal, parts and it is the general purpose of this chapter to consider the reality behind this conception and its bearing on our understanding of the evolution of these plants.

Let us begin with two general observations. First, it is important to realize that the division into monocotyledons and dicotyledons is not merely the expression of the two possibilities of a simple alternative such as is found commonly enough in classifications made for purely analytical purposes. In some Gymnosperms the number of seed-leaves in the embryo is more than two, and therefore the fact that in the Flowering Plants the number two is but rarely, if ever, exceeded becomes a characteristic of positive significance. Second, this character of seed-leaf number does not stand alone and unrelated to other features. Indeed, the fact that it provides so admirable a means of practical classification is in itself sufficient evidence of this, because the division of so huge a group of organisms on any arbitrary, isolated character would assuredly make any further rational subdivision almost impossible, and this is not the case with the Flowering Plants. On the contrary the difference of cotyledon number is accompanied, to a remarkable degree, by correlated features, so that the individual members of the two are in general readily recognizable at sight. If this were not so the criterion of seed-leaf number, being one which cannot be directly detected externally, would be of little practical value.

The Morphological Differences between Monocotyledons and Dicotyledons

The better-known differences between monocotyledons and dicotyledons number about a dozen, but it is not easy to give a satisfactory brief account of them because they vary among themselves, not only in degree of absoluteness and importance, but also in their independence of one another. Taking this into consideration it is reasonable to consider four of them as primary and more independent and the rest as more obviously consequential on one or other of these four. All may then be expressed briefly as follows:

Primary differences
  1. Number of cotyledons
  2. Vascular stem anatomy
  3. Leaf form and venation
  4. Numbers of floral parts
Other differences
  5. Absence of persistent secondary meristems in monocotyledons
  6. Absence of tap roots in monocotyledons
  7. Presence of a scutellum in monocotyledons
  8. Size of embryos
  9. Leaf insertion
  10. Position of the prophyllum.

Before dealing with these, two further points must be made. First, this is not the total of differences, though it is perhaps the total of those which can usefully be discussed objectively, and it is of great interest that there are, in addition, various less-evident distinctions which, although not susceptible to the same kind of description, are nevertheless of considerable significance. Something will be said of these in due course. Second, it must be emphasized that of all the characteristics listed no single one, as there expressed, is absolute. No single character completely and totally separates the whole of one series from the whole of the other, so that however different any one monocotyledon may be from any one dicotyledon this is due, not to any single peculiarity of either, but to a summation of more or less diagnostic features. The most typical monocotyledons (those most distinct from dicotyledons) are plants in which this summation is, in terms of the above summary, ten or nearly.

1. *Number of cotyledons*

This has received particular attention, partly because of the inherent interest of the existence of so seemingly fortuitous a difference and partly because some understanding of it promises to throw light on the vexed problem of the relationship between the two series.

Investigation shows that the presence of a single cotyledon without any hint of bifurcation or duplication is one of the most consistent of all the distinctive characters, but there are several points to be born in mind in coming to this conclusion. For one thing this is the actual diagnostic character of the series and so, in theory at any rate, it would be a contradiction in terms to speak of a dicotyledon with one seed-leaf and of a monocotyledon with two, but in practice allowance is made for these possibilities by taking into account the totality of characters mentioned and treating such cases as anomalies. As regards cotyledon-number these anomalies are surprisingly few, which means that only very rarely have dicotyledons only one cotyledon or monocotyledons two. The latter condition seems, indeed, to be unrecorded, but there are various examples of dicotyledons in which the cotyledons appear to be completely or partially single, presumably through fusion. *Ranunculus ficaria*, *Corydalis cava*, and *Carum bulbocastanum* are good instances and there is a similar state of affairs in some Nymphaeaceae. In *Nelumbium* it is only after a time that the single initial cotyledonary mass becomes recognisably divided into two; in *Eranthis*, *Podophyllum* and *Delphinium nudicaule* the petioles of the cotyledons are united into a sheath through the base of which the plumule duly emerges: in Gesneriaceae the two cotyledons are often very unequal; and in some Piperaceae one of the two has an absorptive function. In certain families the cotyledons may be short and small or even minute, but this perhaps relates more to the condition, well known in both series, where the embryos are very simple and undifferentiated.

The mention of a petiolar cotyledonary sheath is of special interest, not only because it explains why it is sometimes said that while the single cotyledon in monocotyledons is terminal and the shoot lateral, in dicotyledons the opposite is the case, but also because it leads to a further associated difference which may well be of greater significance than appears at first sight. The primordium of the monocotyledon leaf, arising at one point of the circumference of the growing point, quickly spreads round it so that the leaf-base becomes annular, and any structure subsequently differentiated immediately above it must also in a sense be *within* it. Such a position is occupied by the plumule in respect of the single cotyledon and since the former must free itself from the latter the most direct way of doing so is for it to pierce the sheath in what may seem to be, if only for a time, growth in a lateral direction.

It is unlikely that the leaves of plants in which the leaf base becomes annular in this way can ever be exactly opposite one another, and this is perhaps why this type of leaf insertion is rare in monocotyledons. It may also be observed that if, because of this, the leaves have to appear in succession, the condition of true monocotyledony is much easier to understand.

2. *Vascular anatomy of the stem*

The first problem here is to try to distinguish between cause and effect. The usual statement is that while dicotyledons have, in the stem, a limited number of open bundles arranged in a ring, the monocotyledons have many more bundles which are closed and scattered. Here three components are clearly involved, namely the number of bundles, the size of bundles and the arrangement of bundles, and these are plainly related. Put in one way the essential difference is between few large bundles and many small bundles, and this can be taken a step further by pointing out that the possession of a fascicular cambium permits a bundle to be of almost any size within limits, so that the lack of this in closed bundles must almost inevitably be redressed, if comparable quantity is to be obtained, by an increase in numbers. As to the position of the bundles it is easy enough to understand that what is called "normal" secondary thickening could scarcely occur unless the bundles were in a more or less simple and defined ring, though the converse, why closed bundles in solid stems should not also occur in a ring is more obscure. At all events it would seem that the crucial point here, both as regards the bundles in particular and the monocotyledons and dicotyledons in general, is the presence or virtual absence of a functional cambium. This distinction is in fact far from clear cut, though in a very interesting and suggestive way, namely that the not infrequent anomalies are nearly all among dicotyledons. Thus a number of these, especially in the Ranales, have scattered stem bundles, while among monocotyledons a ring of bundles seems to occur only where it is inevitable because the stems are hollow. Again, the bundles of monocotyledons, though they may show traces of cambial cells, are rarely, if ever, truly open, and such methods of secondary thickening as these plants possess are quite different in principle from those habitual in the dicotyledons.

All this suggests that the real criterion here is the presence or absence of a persistent fascicular cambium, and this may in its turn reflect some fundamental difference in the ontogeny of the vascular system. Various other of the distinctions between mono-cotyledons and dicotyledons may certainly be attributed wholly or in part to this absence of a functional cambium and it is not im-possible that, in a broad sense, all the differences may be based on this. However, even if this is so it does not tell us how it may have come about.

3. Leaf form and venation

In one way this is the best-known difference between the two series because it is the most obvious, and indeed the only, visual criterion which is apparent throughout the active life of the plants. The usual statement is that in dicotyledons the leaves are broader and have pinnate or palmate net-venation and that in monocotyle-dons they are narrower and have parallel venation. What is the truth of this and what is its possible explanation?

With regard to the first this difference happens to be unusually obvious among the plants of more temperate latitudes and has thus assumed a somewhat exaggerated importance, for when tropical monocotyledons are comprehended the exceptions are both numerous and notable, including the Palms, most of the Aroids, the Bananas, and several other groups. Conversely there are relatively few dicotyledons with leaves closely akin to the accepted "monocotyledon-type", and the genus *Plantago* seems to be the largest whole group, though there are plenty of other instances, more especially among the Ericales.

A more generally accurate statement would be that compound leaves and lobed leaves are comparatively rare in monocotyledons, and it may even be held that some at least of those which do exist, as in the palms, are not wholly homologous with those of dicotyledons. This is a useful point because it leads on to the possible explanation of these differences, that of the greatest interest here being that the monocotyledon type of leaf, with its parallel sides, parallel venation and indefinite basal growth, is in fact not a true leaf blade but a phyllode, representing only the petiolar part of the dicotyledon leaf. If this is so then most monocotyledons must be regarded as lacking a leaf blade.

Perhaps the most significant difference between the foliar organs of the two series is that the typical monocotyledon leaf often has a more or less circumferential attachment to its axis and indefinite growth from an intercalary basal meristem, while the typical dicotyledon leaf has a localized attachment to its axis and no intercalary meristem, so that it does not, after it has once expanded, increase further in size. It thus appears that the monocotyledon leaf lacks a blade but has an intercalary meristem and that the dicotyledon leaf lacks an intercalary meristem but has a blade. How far either one may represent a compensation for the absence of the other is an interesting speculation.

4. *Number of floral parts*

In all but a few monocotyledons the number of parts in each whorl of the flower is three or some multiple of three (trimerous): in dicotyledons it is normally either four or five (tetramerous or pentamerous). These numbers are to be regarded as basic in the sense that they apply particularly to the complete flower, which is one having calyx, corolla, andrecium and gynecium, all well developed. In both series there are also to be found flowers of a simpler organization with less than these numbers of parts. It is important to remember this because it may explain the major exception to the rule, which is the characteristic and general occurrence of trimery in the dicotyledonous families Polygonaceae, Annonaceae, Myristicaceae and several others. Monocotyledon flowers with a basic number greater than three are a good deal rarer, the best-known being some Aroids, *Potamogeton* and *Paris*.

Perhaps the most interesting point about this characteristic is that it seems so gratuitous, a distinction as it were without a difference. It is hard to see how the function of the flower can be affected by it and there are no obvious indirect consequences of it. Were the respective basic numbers in the two series three and four, these might be related to the cotyledons by pointing out that while the former is a prime the latter is a multiple of two, but in fact the prime number five is commoner among dicots than the number four.

5. *Absence of persistent secondary meristems in monocotyledons*

One particular aspect of this, the absence of a fascicular cambium, has been sufficiently discussed above and we may consider here the rather broader issue of the general relative absence of cambial tissues in the monocotyledons.

That this has not been ranked among the primary differences is because it is even less than those an absolute character, though this is chiefly because of the generalized form of the proposition. Vestiges of cambial activity, both in stem and leaves, are said to have been found in the bundles in nearly all groups of monocotyledons, and there are many dicotyledons in which there is little or no cambial activity, as, for instance, in many water plants and annuals. In short, although it is not true to say that all monocotyledons are without secondary meristems, or that all dicotyledons have them, it does seem true that no plant normally classified among the monocotyledons has a cambium which is exactly similar in position and mode of action to those typical of dicotyledons. Less strictly, monocotyledons are either without secondary thickening or have a special type of it.

This being so two propositions seem irrefutable. The less important perhaps from our present point of view is that there appears to be complete correlation between the presence of a "typical" cambium and the other characteristics diagnostic of dicotyledons. The more profound is the conclusion that the monocotyledons must, by the nature of the case, be debarred from those structural potentialities which are consequential upon the possession of typical dicotyledon secondary thickening. What these amount to is best judged from Chapters 4, 5 and 6, but it may be said here that, in the ordinary sense of the term, the woody habit is the particular perquisite of the dicotyledons.

At the same time it is of the highest evolutionary interest that the Monocotyledons do in fact contain quite a number of woody plants of their own characteristic anatomical design, and indeed what, to many minds, are among the most impressive of all flowering plants, the Palms, are of this sort, and there are plenty of less striking examples. In most of these the outstanding feature is a columnar stem, but in some, *Cleistoyucca arborescens* for instance, there is a kind and degree of deliquescent branching which results in a silhouette very like that of many of the larger dicotyledons. Rather paradoxically annuals are more frequent among the dicotyledons.

What has been said of stelar cambium applies in general also to cork cambium. Bark in the ordinary sense of the term is a tissue or system which is self-augmenting in concert with the secondary growth of the axis and thus capable of providing the more robust external protective covering which large persistent aerial stems need. It is therefore only to be expected that, in this sense, bark is a feature of dicotyledons only and it is interesting to note that these monocotyledons which have their own methods of radial enlargement generally have also their own types of external protective tissues.

6. *Absence of tap roots in monocotyledons*

This again would seem to be connected with the absence of typical secondary thickening, and the circumstances relating to it are very similar, namely the replacement of the more general by something more particular. Thus, while above ground the lack of an axile cambium is made good by special methods of radial augmentation, so in the root system the lack of a master root is balanced by the copious development of minor, fibrous roots, often of a quite remarkable mechanical efficiency. Even some of the largest Palms, which in this respect might be expected to show considerable instability, are securely anchored by this means.

7. *Presence of a scutellum in monocotyledons*
and
8. *Difference in size of embryos*

It seems justifiable to consider these together as different expressions of, or at least as related to, the fact that exalbuminous seeds are more characteristic of dicotyledons than of monocotyledons, which means that the reserve food of the embryo in dicotyledons is more often within it than around it. It would be idle to speculate on the ultimate reason for this, but it is a fair summary of the position to say that while in the monocotyledons there is often a more or less specialized method by which the embryo absorbs from an outside food store, it is, in dicotyledons, as common to find this reserve food actually within the embryo itself. This being so it is perhaps not surprising that the embryos of dicotyledons should on the whole be larger than those of monocotyledons, but in fact it may be doubted whether, if only the albuminous seeds of the two series are compared, there is any real validity in this criterion.

A more pregnant consideration is whether the presence of large exalbuminous embryos in dicotyledons relates at all to the fact that they have two cotyledons, a character, which *ipso facto* would seem to argue the possession of a greater potentiality for food storage. If so, then any difference in embryo size becomes consequential on cotyledon number.

9. *Difference of leaf insertion*

This embodies the fact that while in monocotyledons the leaves are normally alternate or scattered and but rarely opposite, in dicotyledons opposite leaves are general and perhaps predominate. This is an unsatisfactory difference in one way because of the difficulty of determining whether apparently opposite leaves are always truly so in origin, and in the not infrequent cases of so-called opposite leaves in certain monocotyledon families (especially in more specialized aquatic groups) it may be that the leaves are not precisely opposite but alternately very close and very distant from one another, and in fact it is hard to see how leaves with fully annular insertion can ever be strictly opposite. If this is indeed the case then it becomes true that opposite leaves are found only in dicotyledons, and if so then it may well be that the distinction is but another reflection of the difference in cotyledon number.

10. *Position of the prophyllum*

The prophyllum, which is defined as the first bract of a flower cluster, is said to be solitary and posterior in monocotyledons, and double opposite and lateral in dicotyledons. All that need be said here is that if the point is a good one, it is presumably yet a further expression of the difference in cotyledon number and plan of leaf insertion.

The foregoing are the most notable formal differences between monocotyledons and dicotyledons, and are therefore the most important characters on which a diagnosis can be based, but they are clearly not the whole story of the dissimilarities between the two. The above tabulation of criteria is certainly enough to afford a detailed definition of each of the two series but it is not enough to give a full and realistic picture of the plants concerned, and of their total "unlikenesses". It may be argued that this is simply due to the fact that the above are only the leading items in a longer list, and this is true up to a point, but only up to a point. The list can be extended, but when this is done it soon becomes obvious that the nature of the characters involved is changing. Not only are they, as might be expected, less and less absolute, and therefore less diagnostic, but their verbal expression becomes more and more difficult. This can be illustrated familiarly enough.

It is, botanically, an easy matter to distinguish between mono-cotyledons and dicotyledons at sight, and no phanerogamist of experience is likely, in sorting plants, to meet any serious difficulty in deciding, in this way, to which of the two series any individual belongs. There may be occasional problems but these will be rare, and probably due to some special circumstance which obscures the real situation. The important point is that the botanist does *not*, in fact, sort his specimens by the conscious application of the criteria set out above, though some must be involved, but by the use of his experience, which has taught him that plants of one kind of general appearance are almost invariably monocotyledons while those of another are dicotyledons. Each series has its own characteristic facies and this is not solely the total expression of the differences already listed: there is something more. For instance, it may be claimed that monocotyledons are "betrayed" by their foliage and, especially, by the absence of leaf blades, but in fact it is not difficult to recognize monocotyledons even when they have dicotyledon-like leaves, or *vice versa*. Some of this is admittedly due to experience acquired from the traditional taxonomy of books, by which one is prepared to find monocotyledons of such and such a kind, but even this is not the whole story, and it is interesting to speculate how far anyone unversed in botany could separate monocotyledons and dicotyledons by sight alone.

The further this line of thought is pursued the clearer it becomes that the more formal diagnostic characters are accompanied by others, less general and less definable, but important in their contribution to the typical facies of the series concerned. Three examples will probably emphasize this better than any further verbal exposition.

The first concerns homochlamydy. This is the condition in which the perianth consists of only one kind of constituent leaf, or tepal, but this may be due to one or other of two circumstances. The tepals may be all alike because there is no discernible difference between the sepals and petals, or it may be because they are in fact either all sepals or all petals, and it is noteworthy that the former is as characteristic among monocotyledons as the latter is among dicotyledons. Moreover, homochlamydy is much more prevalent in monocotyledons than in dicotyledons, though in neither series is it so widespread as to rank with the distinctions listed above. There is in short a monocotyledon type of homochlamydy and a dicotyledon type and the experienced botanist presented with detached flowers of this kind would have little difficulty in saying to which series each belonged. Whether tepals in monocotyledons tend to be alike for the same reason as the leaves, namely that they are "phyllodinous" in origin, is at least an interesting thought.

The second example concerns hairiness. There are plenty of hairy monocotyledons, and the character has little diagnostic value, but hairiness is immeasurably *more* a feature of the dicotyledons. Indeed, and here is the real point, the comparative hairlessness of monocotyledons is one of their leading indications. Experience teaches us that hairy plants are less often monocotyledons.

The third example is perhaps the most helpful, though it is also the most difficult. The fact is that when all the more definite characters are exhausted monocotyledons and dicotyledons remain distinct in the botanist's consciousness. It may be that it is impossible to put this final distinction into words at all, but this may, at least, be attempted. Is there any one general reason, additional to those already mentioned, why monocotyledons and dicotyledons should look, on the whole, so different? One may be suggested, namely that between monocotyledons and dicotyledons there is a basic dissimilarity of general form which can be expressed only by saying that the former are "streamlined" in a way in which the latter are not. Associated with this is a sleekness and pallor which is commonly unmistakable. Together, these are, it may be urged, some of the most real characters of the monocotyledons.

But what is the status of these last items? Are they fundamental or are they, too, consequential? The streamlining of monocotyledons would seem to owe a good deal to the absence of broad leaf blades, and also to the less copious branching which seems to be associated with the general lack of cambial tissue. Hairiness, as well as what have been called sleekness and pallor, may also be related to leaf form in so far as much of the characteristic dicotyledon hairiness and depth of colour derives from their leaf blades.

This discussion of the criteria in distinguishing between mono-cotyledons and dicotyledons may be summarized as follows:

1. Monocotyledons and dicotyledons are generally so distinctive in appearance that there is seldom any doubt as to which series any individual plant belongs.

2. At the same time there is no single easily describable character which infallibly separates the two.

3. Nevertheless there are certain prominent characters which are specially correlated with the differences between the two, and one of the most nearly absolute of these, the cotyledon number, is used to diagnose and name them. This character is not directly discernible externally.

4. Other leading differences are in the anatomy of the stem; the structure of the leaves; and the numbers of parts in

the outer floral whorls. The two latter are directly discernible externally.

5. Two other familiar, but less profound differences are the relative frequency of persistent secondary meristems; and the size of embryos. These are probably to be regarded as consequential differences.

6. Besides all these there are other differences which, though scarcely amenable to verbal description, contribute to the building up of what has come to be regarded as the typical facies of each of the series. How far these differences also are consequential is not so clear.

7. The most nearly absolute characters appear to be (a) the absence of functional fascicular cambium and the presence of a limiting bundle sheath in monocotyledons, and (b) the more annular leaf primordium in monocotyledons.

8. With regard to most, and perhaps all, characters, anomalies are to be found in both series. There is a strong impression that there are more often "monocotyledon" characteristics among dicotyledons than *vice versa*.

9. There can be little doubt that the difference between them which is, and has been, the most profound in kind and effect, is the absence in monocotyledons of annular secondary thickening by means of a single fascicular and interfascicular cambial ring.

10. This difference in cambial activity, and the nature of the other characters mentioned, suggest that the monocotyledons are, fundamentally, flowering plants which lack certain of the more important organizational potentialities of the dicotyledons, but that, perhaps because of this, certain other particular morphological traits are more fully expressed in monocotyledons than in dicotyledons.

There are two further directions, by no means unrelated to those just discussed, in which a comparison of the two series is of great interest. These are the relative size and importance of the two, and their geographical distributions (see p. 50).

The Numbers of Species in Monocotyledons and Dicotyledons

The criterion of size can be applied in several ways to a group of organisms, the choice depending on the purpose involved, but the usual criterion is that of taxonomic classification, which expresses itself in numbers of species and other taxa. On this basis the two series appear to be highly disparate, there being about four times as many species in the dicotyledons as in the monocotyledons, and for this reason the former series is commonly described as being four times as large as the latter. This statement is, however, an over-simplification and it is important to try to arrive at some conclusion as to how true it really is.

With regard to the actual taxonomic numerical relation it is to be borne in mind that the ratio of species is reflected pretty closely also in the two larger categories of genera and families, and that in consequence these taxa are of about the same average size in both series. In particular the average number of species *per* family is about the same, some 600. At first sight this would appear to reinforce the impression of disparity but there are other considerations to be taken into account.

As will be made clearer in the next chapter the use of the term *species* is only a way of expressing in simplified form some of the peculiarities of nature, and the facts underlying its use are the existence of a vast number of separate individual plants, no two of which are precisely alike in all respects. This being so it would plainly be a nearer approach to truth to express the size of groups in numbers of individuals rather than numbers of species. Unfortunately there is no practicable way of doing this, but comments upon it are not entirely without value. On several grounds it would seem unlikely that there is, in the numbers of individuals, any discrepancy between monocotyledons and dicotyledons comparable with that of their species numbers. Certainly, dicotyledons are, for reasons specially associated with their structure and growth-forms, on the general average larger plants than monocotyledons and seem to predominate in many familiar kinds of vegetation, but it is also to be remembered that the larger the individuals the smaller their numbers are likely to be, as is easily illustrated by the fact that, in an oak-wood, bluebell plants may be more numerous than oak trees. Again, the comparative infrequency of monocotyledons among the larger plants of the world may well be compensated, or even more than compensated, for by their great numbers in such vegetation as grassland and in many aquatic habitats. At least it would seem rash to claim, on any basis of individual numbers, that the dicotyledons are, by any considerable margin, a larger series than the monocotyledons.

On the question of species and their value it will be evident that any expression of size in terms of species depends, for any reality it may possess, on the assumption that the average difference between species is the same in both monocotyledons and dicotyledons, and here again there are one or two important points to be remembered. Classification of the anthropogenic sort used in biology, rests ultimately on the fact that, and is only possible because, there is an essential unevenness of likeness between living individuals, this unevenness being distributed in such a way that in almost all circumstances it is possible to recognize certain individuals as being more like one another than they are like any other individuals. Indeed this is one of the most satisfactory ways of putting into words what is commonly meant by a species, it being a collection of individuals of which this remark is true. In other words there are in nature structural and other discontinuities or gaps of almost every dimension within any major range of form as a whole such as the monocotyledons and dicotyledons present, and it is first the recognition and then the valuation of these discontinuities which is more than anything else the real work of the taxonomists. Thus in the conception of the species there are two ingredients, the similarities of the individuals within the species to one another and the gulf which separates all or any one of these from the individuals of another species. There is no *prima facie* reason for supposing that the differences between the *individuals* in the monocotyledons are much larger or smaller than those between individuals in the dicotyledons, and this would suggest that it is the other ingredient which is most likely to be concerned in the numerical disparity between the two series, namely the values of the gulfs between taxa.

But the assumption that the species of the two series can be accepted as of reasonably equal taxonomic value, i.e. as being equivalently unlike in form, can scarcely be denied. The species is itself different from other taxonomic categories because it is (in the artificial framework of taxonomy) an ultimate division, and is therefore not affected by those variations of content which are inherent in genera, which may consist of any number of species, or families, which may consist of any number of genera. Moreover, its criteria are still peculiarly those of direct sensory appreciation, and since the characters in diagnosis are much the same in both series it would not seem very likely that there is any great difference between them in the amplitudes of their specific differentiation. With larger categories the case is rather different because the wider the discontinuities between taxa the less easy it is to estimate their relative values. It may be comparatively simple to argue that the difference between one pair of species is comparable with the difference between another pair, but much more difficult to do so for many of the pairs of families, in which so many more points of distinction must be taken into account. It would therefore appear true to say that there are greater variations or extremes of difference between families than between genera within any family, and between genera than between any species within a genus. If this is so, it follows that although the average gulf between species may be much the same in monocotyledons and dicotyledons the average gulf between the higher taxa, and especially between families, may be very different in the two series.

This brings the argument to a very important point. We have seen that the size of organic groups may be expressed from the taxonomic point of view as species numbers, or from a more absolute point of view as the number of individuals, but to these must be added a third method, especially relevant from the point of view of evolution. In so far as evolution is gradual change with time and in so far as species represent comparable quanta of change from a single origin, the total number of species in a group is some measure of the amount of evolution which has gone to the building up of the group as a whole. But evolution can also, and just as significantly, be measured or assessed by the total extreme divergences of form which it has produced, and it is in this respect that a comparison of the monocotyledons and dicotyledons is perhaps most interesting.

The difficulty of making any exact comparative assessment of structural divergence in living things needs no emphasis, but if we consider, for a brief moment, such diverse monocotyledons as Palms, Screw-pines, Orchids, Yams, Gingers, Aroids, Grasses, Duckweeds, and some of the more bizarre Saprophytes, it is certainly permissible to wonder whether any very much greater range of design is to be found within the dicotyledons. True, as we shall see in Chapters 5 and 6, monocotyledons are much less multifarious than dicotyledons, and it is also true that their morphological differences impose on the monocotyledons certain broad restrictions of organization, but it may nevertheless be doubted whether they do in fact show much less total divergence of form than the dicotyledons. At least there is an astonishing set of parallels between the two series in which many of the most specialized groups of each are reflected (see Chapter 12).

To sum up, although the number of species of monocotyledons, is only about a quarter of the number of species of dicotyledons, it seems far from certain that this is a fair expression of the relative sizes of these two groups in many important respects, notably those of individual numbers and total range of structure. It seems more likely that because the monocotyledons lack certain structural potentialities which the dicotyledons possess, they lack also the contribution these make to the body taxonomic, but that this has seriously restricted their total breadth of expression, as opposed to their mere multiplicity, is more doubtful.

What may seem, on this view, to be an inconsistency between species number and total range of form is perhaps, then, to be explained on the lines that, while the species in the two series are more or less the same in taxonomic amplitude, the larger taxa of the monocotyledons are more isolated from one another, both structurally and phyletically, than those of the dicotyledons, and that the greater species numbers of the latter are provided by inter-mediate conditions more continuously joining some of the extreme types.

The situation thus suggested can be pictured in a number of analogies, of which three may be mentioned. In the simplest the monocotyledons and dicotyledons may be thought of as two instrumental keyboards similar in length but having in the case of the latter about four times as many notes, these being separated by smaller tonal values. In a second analogy the two series may be pictured as nets of equal area, in one of which, the monocotyledons, the meshes are larger and the knots fewer, and in the other of which, the dicotyledons, the meshes are smaller and the knots more numerous. But, as will be seen at greater length in the next chapter, there is a third analogy much more complete than either of these in what is there called the *stellar analogy*. In this the two series are thought of as twin constellations of approximately the same overall dimensions. In the monocotyledon constellation the constituent stars, which are of many sizes, are fewer and, on the average, further apart; in the dicotyledon constellation the stars are much more numerous and, on the average, closer together.

Because the time factor involved in speciation is inherent in evolutionary assessments these problems of the "size" of monocotyledons and dicotyledons have a direct bearing on their histories. If the species in the two series are accepted as approximately of the same taxonomic value, then the numerical disparity between them is capable of historical explanation in one or other of three ways. One is that the monocotyledons and dicotyledons have had the same length of history but that the production of species has been much more rapid (about four times so) in the latter. The second is that speciation has been at the same rate in both, but that the history of the monocotyledons is shorter than (about one quarter) that of the dicotyledons. The third is that speciation has been at the same rate and for the same length of time in both but that only a small proportion (about one in four) of the monocotyledon species have survived.

It is not easy to see what positive evidence could be adduced in support of the last possibility short of the discovery of the remains of great numbers of extinct monocotyledon species. If on the other hand non-survival means that most of the new forms produced by monocotyledons have been non-viable or lethal, this would merely amount to the lesser frequency of speciation envisaged in the first of the three contingencies. Moreover, any such suggestion as this would attribute to the monocotyledon lineage in general such a degree of ineffectiveness as would not only be foreign to all experience but would almost certainly have expressed itself in recognizable peculiarities of group-constitution and classification. It thus seems a more difficult explanation than either of the others.

These others need some further consideration. With regard to the first the simplest conception would be that monocotyledons and dicotyledons are groups of equal age, having had a contemporary origin, and that each has since pursued its own independent evolution, but this is clearly not the only possibility. Another is that subsequent to a contemporary origin there has been repeated reinforcement of one stock from the other, which, more closely defined, means either the repeated appearance of monocotyledons from a dicotyledon source or *vice versa*. There might even have been a reciprocal production of new forms, and it is interesting that this seems seldom, if ever, to have been suggested. The second possibility, that the monocotyledons have had a shorter existence than the dicotyledons, is perhaps the simplest explanation of taxonomic disparity and is one of the most important foundations for the belief that monocotyledons have been derived from dicotyledons.

As between monophylesis and polyphylesis there is one very important consideration. If the two series have been distinct *ab initio* it is permissible to surmise that their speciation has been, in a large sense, continuous and, presumably, though not necessarily, at a constant rate, fluctuating only under such influences as might be expected to affect both series simultaneously and comparably. If, however, one series is polyphyletically derived from the other, which has thus reinforced it at two or more intervals of time, it would seem that its speciation is likely to have been less constant, and rather to have been augmented sharply at intervals by sudden accessions corresponding to these reinforcements. *A fortiori*, if there has been reciprocal reinforcement constancy of speciation can still less be expected and numerical disparity between the groups would be even less significant and capable of interpretation, since it would depend, among other things, on the degree and frequency of reciprocation.

In short, one reason for taxonomic disparity may be that the opportunities for parity have been less on one side than the other, meaning that the species-producing stock has been more restricted in some way in the one series than in the other. If the two sides are truly of equal age and autogeny, this restriction can only have expressed itself during their subsequent histories, and can only be indicative of a differential rate of speciation taking one or other of two forms, a slowing down of a continuous function or discontinuity in a function of given rate. If on the other hand the two stocks are of different ages and of closer relation because one has been derived from the other, some numerical disparity would be inevitable to such a degree, with or without constancy of speciation, that its absence would be more remarkable than its occurrence. Similarly, if the derivation of the two series has been reciprocal it is almost inconceivable that there should not be disparity.

The Geographical Ranges of Monocotyledons and Dicotyledons

A geographical comparison of the monocotyledons and dicotyledons shows that there is little or nothing to choose between them. Both are completely world-wide in range, in that they are found wherever there is flowering plant vegetation, and broadly there is a most striking regularity between their respective distributions, exemplified by the fact that floras everywhere contain both monocotyledon and dicotyledon species in about the same *proportion** while there are no floras, even in the most isolated parts of the world which consist, to any very large degree, of members of one or other only. Both are found as far north as there is land, and of the only two families represented in Antarctica one belongs to the Monocotyledons and one to the Dicotyledons. Much the same is true ecologically, and although arboreal vegetation naturally tends to include more dicotyledons, some other types, such as grassland, contain more monocotyledons. Again, taking all in all, it may be said that it is a unit of the smaller series, the family Gramineae, which is the most completely distributed in the Flowering Plants. On the other hand the wide distribution of the monocotyledons seems more to be a matter of only a few groups than it is in the dicotyledons. Of the ten most widespread families, four, Gramineae, Cyperaceae, Orchidaceae and Liliaceae are monocotyledonous, but to the thirty widest families the monocotyledons add only one more, the Juncaceae. There are also, it is true, some very small families of aquatic monocotyledons that are very widespread, but their ecological restriction makes them hardly comparable. It may perhaps be added that many of these facts are even more interesting when it is remembered that the monocotyledons are, to a notable extent, without the special dispersal mechanisms to which wide distributions in the dicotyledons are often in so facile a way attributed. No doubt there are a few very minor geographical differences but these are only enough to substantiate the general statement that whatever the disparity between monocotyledons and dicotyledons in species number it is not accompanied by any comparable difference of distribution.

---

* It should be noted that this ratio of occurrence has no direct or necessary relation to the proportion of species in the two series: theoretically it could occur whatever these might be.

This leaves a situation of great interest. There are two sorts of flowering plants, one kind being *in certain respects* four times as considerable as the other, but the total distributions of the two are virtually alike, while everywhere a ratio of about one to three in species present is more or less closely retained. It would seem at first sight that such a simple pattern of affairs must surely point to some equally simple historical explanation, but it is difficult to see what this can be. Whichever of the three phyletic possibilities outlined above may have prevailed, the present geographical condition might have arisen, and it seems impossible to make a choice on this basis. What we lack is any knowledge of the relative rates of spread of different flowering plant types, or to put it rather differently, of how long any particular group has had a world-wide range, and until there is some information on this matter it seems idle to speculate further. It is, however, justifiable to express the opinion that the explanation which least strains the imagination is that the two series are of equally remote origin but that for reasons not at once self-evident, the evolutionary elaboration or virtuosity of the dicotyledons has been much more copious, this being expressed in one way by the number of species. Considering all that was said in the earlier part of this chapter it would seem most likely that the reason, obscure as it may be, has something to do with the morphological differences between the monocotyledons and the dicotyledons.

Any book which deals with the historical development of the Flowering Plants might be expected to include an early chapter comparing and contrasting monocotyledons and dicotyledons, because the existence of these two parallel series is the most obvious evolutionary problem in the whole phylum, but the survey that has just been given here fulfils an additional and more particular purpose, that of providing a simple overture to the main theme of this book. This, as the Introduction explained, is the presentation of a small selection from the immense store of relevant facts that these plants afford and the testing against them of some of the many conceptions which go to make up the "theory of evolution" as that is commonly understood today. It is, above all, concerned with trying to determine the degree of validity of certain evolutionary hypotheses, and there can scarcely be a better simple example of the testing nature of the Flowering Plants when used in this way than the matter which has been the subject of this chapter.

NOTE ON LITERATURE

Further details of the comparative anatomy of the monocotyledons and dicotyledons may be sought in Solereder, H., *Systematic Anatomy of the Dicotyledons*, English trans. Oxford, 1908; in Solereder, H. and Meyer, F. J., *Systematische Anatomie der Monokotyledonen*, Berlin, 1928–33; in Metcalfe, C. R. and Chalk, L., *Anatomy of the Dicotyledons*, Oxford, 1950, and in textbooks such as Eames, A. J. and MacDaniels, L. H., *An Introduction to Plant Anatomy*, 2nd ed., New York and London, 1947, and Esau, K., *Plant Anatomy*, New York, 1953.

Useful summaries of the differences between the two series are to be found in Rendle, A. B., *The Classification of Flowering Plants*, Cambridge, 1925 and 1930; in Wettstein, R., *Handbuch der Systematischen Botanik*, 3rd ed., vol. II, Leipzig and Vienna, 1924; and in Lotsy, J. P., *Vorträge über Botanische Stammesgeschichte*, vol. III, pt. I., Jena, 1911. Much interesting information, especially relating to the monocotyledons, is to be found in Arber, A., *Monocotyledons*, Cambridge, 1925 and in Priestley, J. H. and Scott, L. I., *An Introduction to Botany*, 2nd ed., London, 1954.

Much of the more scattered previous literature is summarized by Bancroft, N. in "A review of literature concerning the evolution of monocotyledons," *New Phytologist*, **13**, 1914, and there is a more recent return to the subject by Hill, A. W. in "The monocotyledonous seedlings of certain dicotyledons. With special reference to the Gesneriaceae," *Annals of Botany*, N.S. **2**, 1938. For the phyllode theory see Arber, A., "The phyllode theory of the monocotyledonous leaf . . .", *Annals of Botany*, **32**, 1918.

CHAPTER FOUR

A REVIEW OF THE FLOWERING PLANTS

I. GENERAL

Flowering Plants and their Classification

SINCE every formal classification of the Flowering Plants is, in effect, an abbreviated description of them it might be thought that any one of these would provide a framework for the review of the group which this and the next two chapters comprise, but this is not quite the case. These classifications were shaped, not so much with the object of depicting the Flowering Plants in terms of the evolutionary processes which may have produced them, as with the primary object of making them, for purposes of identification, readily amenable to particulate analysis, and something is needed which will give a truer picture of these plants in terms of the part they play as the fundamental biological ingredients of almost every sort of land and fresh-water environment.

There is another reason why our wagon must not be hitched to any formal classification. Science has made astonishing progress in many directions during the last fifty or a hundred years, but this advance has been uneven, and in certain respects not as great as might seem. There is an impression, which it is becoming increasingly difficult entirely to ignore, that certain kinds of problems, among them some of those encountered in evolution, are intricate in a way that makes it very difficult for the human mind fully to grasp them, and that when we claim to understand them we do in fact comprehend only some simplification of them which we ourselves have invented. Such a method of invention and simplification is often, and with justification, employed, but it must be used with caution and restraint lest the shadow of the diagram becomes identified with the substance of the real picture.

Simplification of this sort is particularly relevant here because the very act of classification is essentially this, and there can scarcely be a better example of it and of its dangers, than that of the man-made classifications of living things. In this instance the harsh necessity for simplification; the nature of the simplification itself; and its consequences, are all apparent. It is the extraordinary variability and multiplicity of the degrees and kinds of relationship between different plants and different animals which is so impossible to deal with faithfully in ordinary language; the remedy adopted is to regard these almost infinitely diverse relationships as all falling within one or other of a severely restricted number of categories (the classificatory hierarchy) in which all the members of any one are assumed to be equivalent in rank; and the consequence, all too familiar today, is that it is almost impossible to think of nature except in the phraseology of this simplified representation.

Biological classification also illustrates the limitations of human intercommunication. For nearly a century now classifications have generally claimed to be *natural*, in that they have essayed to show genealogical relationship as well as, or rather than, mere resemblance, but this is a very difficult thing to do unless one has recourse to some analogy, such as the "genealogical tree", which can be represented either pictorially or by a model.

Unfortunately, a model can only be completely communicated to those who can see and feel it, and even a picture cannot tell more than a certain amount because, even allowing for perspective, it is really only two-dimensional. The obvious remedy for these shortcomings is to supplement, or wholly replace, plastic or pictorial representation by verbal description, and this of course is what is almost invariably done in practice. This has the advantage that it can be amplified to almost any extent, but it has the serious disadvantage that words are only intelligible when arranged in a linear, one-dimensional sequence.

So it is that when a formal classification of a group is consulted there is being consulted something which, although appearing to state the facts adequately, is actually a major simplification of them in two very important respects, their description in terms of certain artificial values and their arrangement in a certain order. In short, it is the habitual practice to describe, and indeed to think, of organisms too much in terms of abstractions called *species, genera, families* and the like which do not in fact exist as such entities in nature. These words are merely verbal counters by which can be expressed in suitably simplified form the overwhelming complexity of nature itself. A treatment so artificial can hardly fail to lead to some lack of understanding and an attempt has therefore been made in these three chapters to describe the Flowering Plants in a way which, it is hoped, may avoid at least some of the distorting effects of a too slavish attachment to these terms.

In doing this there is no intention of denying that simplification of this kind is necessary, for it is certain that without it real progress in biology would be difficult and we should be floundering still among what seem to be the inconsistences and contradictions of nature. For many, but not all, purposes a simplified verbal coinage is, and always has been, a necessity. Nor is anything that has been said intended as a criticism or denigration of the methods of taxonomy as a scientific routine, or of those whose part it is to carry it out. Without the devoted, and often unacknowledged, spade-work of the taxonomists we should remain biologically almost inarticulate.*

Simplification of the kind which has been described is seldom arbitrary but is generally determined by the circumstances of the facts with which it is concerned, and in biological taxonomy it is based on a fundamental fact of nature, namely that while no two organisms are precisely alike, almost any one closely resembles many others. That is to say the differences and similarities between them are not entirely chaotic to our senses but follow a plan or pattern. It is the realization of this plan which, in one direction, has led to that particular simplification which is the framework of modern taxonomy, and in the other towards the belief that the developmental plan of nature has been an evolutionary one.

When we use the words *Flowering Plants* or *Angiosperms* we are rightly speaking in the plural and refer, fundamentally and in reality, to the individual plants which go to make up the huge group to which these names have been given. In fact what we call the Flowering Plants is an immense multitude of individual plants more or less like or unlike one another to almost every degree within the embracing definition of those words. It is the

---

* A brief but useful account of how taxonomy has developed and of the attempts that have been made to improve and rationalize it is given by Heslop-Harrison, J. in *New Concepts in Flowering-Plant Taxonomy*, London, 1953.

existence of this colossal number of separate and, in the ordinary sense of the word, independent plants, which is the basic objective fact of the situation and which provides the tangible material with which the botanist has to deal.

There is no ready means of calculating accurately the total number of individuals in existence at any one moment, but it is worth while emphasizing how vast the number must be. Bearing in mind, first, the size which is likely to be the average for all individual Flowering Plants (and it is worth noting that the average in the more familiar temperate floras is probably considerably less than elsewhere) and, second, the amount of the world's land surface which is to be regarded as virtually devoid of plants, it would seem probable that to base a calculation on an assumption that there are, roughly, as many individuals as there are square feet of land is not to be too greatly in error. There are, as for instance in certain grasslands, situations where there must be many individuals to the square foot. On the other hand there are many places where there are no plants at all, or where the number must be less than one to the square foot, as in the case of large trees, the trunk bases of which may cover considerable space. The total area of land in the world is generally said to be in the neighbourhood of 50,000,000 square miles, and there are nearly 28,000,000 square feet in a square mile, so that the total number of square feet of land and, by our calculation, the total number of individual Angiosperms will be the well-nigh incredible number of 1,350,000,000,000,000, or considerably more than a thousand billion. This total has to be distributed over about a quarter of a million species, so that the average size of the populations which are given the taxonomic rank of species is more than 5,000,000,000 individuals each, a figure not without some interest in comparison with the most authentic corresponding figure in animals, that for the human species, which is said to be something over 2,400,000,000. It is also worth remembering that many flowering plant species undoubtedly consist of far fewer than the average number, some of them indeed of only two or three or a handful of individuals, so that there must be some others at least in which the numbers are much above the average. It may well be of course that the figures should be even higher, but on the other hand if the concentration is diluted to no more than one plant per square yard instead of one per square foot the numbers will be only one figure shorter and still quite astronomical.

These figures help us still better to appreciate the necessity for simplification which has already been mentioned and the fact that the history of plant classification has in considerable degree been the attempt to reduce these almost inconceivable numbers to more manageable and memorable proportions by the subjective process of categorization. That such a process of simplification is possible with any kind of reality (and it *has* a more than negligible measure of reality) is, as has been said, due to the biologically fundamental fact that the individual members of this great multitude show mutual resemblances and differences of a very definite sort, of which the outstanding feature is that while no two individuals are exactly alike most individuals are closely similar to certain other individuals, or to express the same in rather different words, while the differences between the individuals are multifarious they are neither completely chaotic nor completely graduated, the general rule being that *the greater the degree of difference the less frequently it is manifested.* The vast majority of individuals thus have at least some, and more often than not very many, other individuals to which they are extremely similar, or conversely, from which they differ very little.

These considerations point to two ways, doubtless out of many, in which it is useful to picture the Flowering Plants. These with the exception of the epiphytes and some parasites, live rooted in the soil, and thus come to form in total a thin layer of vegetation over the surface of the land, a layer which may, with the qualification just mentioned, be regarded as one individual thick, since no two plants can exactly occupy the same soil space. But the distribution of these individuals is, in all three spatial dimensions, irregular. They may be close to or further away from one another in any horizontal direction, that is to say in dimensions of length and breadth, and in addition, they may vary among themselves in height so that the layer they compose also has a variable of thickness. Thus from one point of view the Flowering Plants as a group may be represented as an excessively and variably thin, variably continuous, layer on the surface of the land. The importance of this representation is that this tenuous layer affords, either directly or indirectly, all the sustenance and nearly all the environmental shelter of the land fauna.

The second way in which the Flowering Plants may be represented in the mind's eye is in accordance with the similarities and differences between individuals. Here the representation is necessarily less objective because there is no system of measurement by which likeness or unlikeness can be assessed with any accuracy, but if we agree to regard the morphological differences between individuals as analogous with the spatial (and therefore measurable) differences between the individuals in the first representation, then something useful can be done. In such case, and with this proviso, the Flowering Plants can be pictured as a vast number of separate entities, the individuals, arranged in a three-dimensional continuum, or, to vary the metaphor, floating in an immense body of air or water, their positions in respect of one another expressing their form relationships and to some indeterminate degree their kinships also.

It should be realized that in a stricter mathematical sense this will be an imperfect picture, for the reason, which has often been pointed out, that plants differ from one another in many distinctive ways, each of which should, theoretically, be incorporated in any perfect representation as a separate dimension. But even if this be an unattainable ideal something considerably short of it has value, and a great deal can be done by thinking of the differences between individuals in a simpler way. They can, for instance, by combining certain kinds of differences, be thought of as so few in number that they can be depicted much more easily, and if, indeed, the number is no more than two or three, they can be accounted for in a two- or three-dimensional plan. Thus if we imagine all the various detailed differences between individuals as amalgamated under not more than three master heads, such for example as life-form, foliage type and flower type, then our mental picture of the Flowering Plants as a vast collection of units floating in the three dimensions of a continuum, with each unit separated from the others in proportion to its degree of difference from them in these respects can be regarded as a comparatively faithful one.

The stellar analogy

Although the picture just drawn must, as far as the Flowering Plants are concerned, be a mental image only, which cannot through lack of knowledge be transferred to real form, there is elsewhere in nature an actual state of affairs which shows a re-markably close resemblance to our imaginative picture. This is the stellar system of the heavens, for the kind of distribution of the individuals in the Flowering Plants which we have portrayed is just the kind of distribution which is characteristic of the stars in the firmament. Astronomy thus affords us a very useful analogy in general between the stars of the sky and the in-dividuals of a biological group or the members of any taxonomic category.

The chief feature of this stellar kind of distribution is that the units concerned are distributed not evenly, but to varying degrees in groups or clusters, in each of which the constituents are much closer to one another than any one of them is to any unit outside the cluster, and this kind of spacing, as we have seen, is the characteristic feature of the morphological relations of living plants and animals. We have also seen that this, in turn, is the foundation of biological classification into such categories as families, genera and species, which are no more than boundaries drawn to inclose certain clusters of individuals. If a cluster of living individuals is so simple that it cannot easily be broken down further into constituent clusters then the word species is usually attached to it: if the cluster is more obviously complex but still fairly simple on the whole then the word genus is applicable; and if it is still more complex and heterogeneous in respect of its constituent clusters then the word family can be applied. Thus a family may be likened to a comparatively large sphere or other outline within which are smaller spheres to which its genera correspond, within each of which again are still smaller spheres to which the species of the genus correspond, and which themselves are each made up of the ultimate real units the individual plants, which, as shown above, are on average vastly numerous.

What has been said about the *stellar analogy* has shown that it has two particular values. It has shown that this analogy can be of real value in any attempt to portray the facts of nature in a way which, while allowing escape from the formal framework of taxo-nomy, at the same time maintains a reasonable degree of contact with it, and since such an attempt is the main purpose of the present chapters this, the *stellar analogy*, will be invoked repeatedly in later pages. It also shows that the task of the taxonomist who essays to produce a truly natural classification has always been, and still remains, that of recognizing what, in the words of the analogy, are the real clusters; that is to say his task is that of drawing subjective boundaries *round* the real clusters and not *through* them, and this in turn means giving to his categorical conceptions the right discriminating characters.

But this by no means exhausts the practical applications of the stellar analogy and it can be extended to several more particular astronomical phenomena which have remarkable parallels in systematic biology. For instance, what we have spoken of as clusters are, by the astronomers, distinguished as being of two sorts, globular clusters and open clusters, the essential difference between these in terms of our analogy being the number and closeness of the stars composing them, those of the latter sort being apparently fewer and more widely spaced. Much the same is the constellation, which has been defined as a rather limited number of relatively conspicuous stars within an imaginary outline. Then there are double stars or pairs which are much closer to one another than they are to any other stars, giving a twinning effect such as is often discernible in taxonomic biology. Of these double stars there is a special kind, the binary stars, in which there is evidence of unusual and real affinity or attraction between the two. There are such simple parallels as giant stars and dwarf stars, which compare inevitably with large and small families and genera. Finally there is the word satellite which, at least in its more general connotation, gives an excellent impression of the nature of the relationship between some taxonomic groups and others.

In addition, there is another interesting parallel between the stellar system and the organic world which, though less factual is nevertheless striking, namely that the more intently and minutely each is studied the greater the number of unities that become apparent. The higher the resolving powers of the telescope the greater the number of clusters that can be seen, and the more exhaustively a group of plants is studied the more small taxonomic units can be descried (*and* described) within it. In the other direction the pictures of both heavens and plant world can be simplified to almost any required degree by, in the one case, reducing the magnification, and in the other by defining fewer and larger groups.

Biological classification

It has been said that it is the task of the taxonomist to use the right discriminating characters for his categorical conceptions, and the taxonomists themselves are likely to be the first to point out that this is little more than a platitude, since rightness is something which can only be determined by reference to circumstances. This is perfectly true and when we say right we mean right in relation to some particular purpose. In the actual context above the purpose is that of delimiting groups of blood relations. In ordinary taxonomy the primary purpose is differential identification, and the characters selected must therefore be of the sort that can be of use in this way. In the study of evolution, which is our purpose here, something different again will be most valuable, and we must now consider what this may be.

We have already seen that, taking the whole organic world into account, the role of the Flowering Plants is the basic one of providing habitats for nearly all the land fauna, as well as for many other plants, and since these other organisms must be in harmony of scale with their habitat, the important features of the Flowering Plants from this point of view which is clearly the evolutionary point of view, must be those which most directly determine the kind of environment they afford. The prime necessities of an environment are a suitable food supply and an appropriate amount of shelter, and both these the Flowering Plants provide in bewildering variety. Shelter, using that word in a wide sense, is the simpler matter, and it must suffice to say that the life-form of a plant is likely to be the chief factor in deciding what sort of habitat it will offer to other biota, because, to mention but two obvious instances, the larger an animal is the larger the bulk of cover it requires, while the relative permanence of sub-aerial plants is clearly of first importance to those animals which pass their lives on or within them. The subject of food is more complicated and properly needs more space than is available here, but in the briefest terms it may be said to have three main aspects, first the materials provided by the vegetative parts of plants, second the products of their ripened seeds and fruits, and third the nectar and pollen of flowers. The first of these is, again, very much a matter of life-form, but the others, and especially the last, are a reflection of something quite different, namely floral form and function. It may be suggested then that the value of any flowering plant as a constituent of a vegetable habitat depends on its life-form and floral organization more than on anything else. These are the principal considerations which lead to biological difference.

It is the object of the presentation of these plants towards which we are working here to sort out the different kinds therein on this biological manner rather than on a purely taxonomic basis, so as to portray the directions which evolution has taken and the progress it has made in them, and we cannot, therefore, do better than to found our analysis on the two general features just mentioned. But we can go one step further than this. Life-form and floral organization are common to all flowering plants, and many kinds of these plants differ from one another only in the quantitative values of certain characters, but in other kinds there is a distinction also of quality or, as it is usually expressed, of specialization. This, though not susceptible to estimation so readily as are the others, can be regarded as a third factor, so that we can describe any given flowering plant not only according to life-form and floral organization but also according to their sort and degree of difference from other flowering plants. It is on this plan that the monocotyledons and dicotyledons will be reviewed in the next two chapters. As a final preparation for this it is desirable to make a more generalized preliminary analysis of their larger taxonomic units, namely the families, partly because these so well illustrate the stellar analogy, and partly because doing so gives a valuable overall introduction to the whole group of the Flowering Plants.

A Brief Review of the Families of the Flowering Plants

The family is the largest in value of the three taxonomic categories in common use, and may be thought of as the largest with any real claim to be regarded as natural, in the sense of being composed entirely of blood relations, because the category above it, the *order*, is very seldom used, partly for the reason that it is generally too large for convenience, and partly because it is (no doubt for this reason) more often than not unblushingly artificial. Thus the families, or some of them at any rate, are the best examples and taxonomic translations of the largest coherent groups of the Flowering Plants. Moreover, the family, coming as it does in the hierarchy above the genus and the species, contains both these in varying proportion and numbers, and so has a variability of content which, from the evolutionary point of view, is of great interest and value. It may also be well to remind the reader here that the criterion of a family is not its mere size but the extent to which its plants differ from other plants. It is essentially a measure of structural isolation, so that, in the words of the stellar analogy, the families represent the larger and more mutually isolated stars and star-clusters in the Angiosperm firmament.

Taxonomists nowadays divide the Flowering Plants into something between 300 and 400 families, according to taste, which means that they consider that there are this number of particularly well-marked and coherent groups of species. Many of these families appear in every classification because everyone agrees as to their validity, as for instance Gramineae, Cruciferae, Labiatae, Compositae and so on. Many others appear equally generally but with some slight variation of content from time to time, their exact boundaries being more subject to personal opinion, as in the Papaveraceae, Liliaceae, Saxifragaceae, Caryophyllaceae and Ericaceae. Others appear only in certain classifications because only some taxonomists regard them as worthy of separate recognition, as for example Zosteraceae, Winteraceae, Chrysobalanaceae, Grossulariaceae and Selaginaceae. Still others are found in only a few classifications or even only in one, because, for a variety of reasons, they are accepted by only a few, or even by only a single botanist. Thus families are not all equally obvious, which is but one way of saying that the distinctions and isolations between them are not all of the same value. Those which are held to be particularly distinctive may be placed in orders of their own, as is the family Casuarinaceae, and this is really the equivalent of giving them promotion in the hierarchy.

Families are made up of genera which in turn are composed of species. A genus may contain only one species, and a family only one genus, so that a family may be composed of only one species. On the other hand the genera may be numbered in hundreds and the species in thousands. This difference in size has one very interesting imponderable. Since the family is recognised as a high-level category, generally with many components, a group containing very few is not so easily recognized as a family. For instance there are many small genera which are probably as isolated structurally from one another as are most families, and which therefore have the same theoretical claim to recognition, but a natural reluctance to give high rank to small units militates against this. This has caused some unbalance in the past and also some limitation in the number of families, and its correction by the erection of more small families is a prominent trend in taxonomy today. Conversely, if one and the same comparatively small isolation is shown by a great number of genera and species, the more impressive it seems and the easier it is to make it the basis of a family. But such considerations threaten to involve us in the thankless task of evaluating differences and the only practicable way of conveying the size of families is by the number of genera and species they contain.

By common consent the largest of all families is that of the Compositae, which are estimated to number at least a thousand genera and about 20,000 species, but recently the numbers of Orchidaceae known have much increased so that today this family with over 700 genera and about 17,500 species is by no means a bad second. Third and fourth, with not much between them, are the Papilionaceae, 440 genera and 10,750 species, and Rubiaceae, 500 genera and 9,000 species. Fifth and sixth are respectively the Gramineae with 600 genera and 7,500 species and the Euphorbiaceae, 300 genera and 5,750 species. The three families Labiatae, Scrophulariaceae and Cyperaceae have more than 4,000 species each: and six, Melastomataceae, Myrtaceae, Asclepiadaceae, Liliaceae, Acanthaceae and Umbelliferae, have over 3,000 each. Some ten others have 2,000 species or more, and from this point the numbers of species diminish till, at the other end of the scale, are families of minimum size, namely with one species only. In such small and isolated groups it is particularly easy to see minor differences between the individuals, but even so the total number of strictly monotypic families approaches 50, or more than 10 per cent of all. The preponderance of smaller families is seen in the following figures, viz.

containing less than 500 species	300 families
with between 500 and 1,000 species	40 families
with between 1,000 and 1,500 species	18 families
with between 1,500 and 2,000 species	14 families
containing more than 2,000 species	20 families

Assuming the total of families as the higher figure of 400 and the total number of species as 250,000, the average size of a family is about 600 species.

But the families of Flowering Plants differ among themselves not only in mere numbers of constituents, which is a purely quantitative disparity, but also in the way these differ among themselves and group together, and these are more qualitative disparities. Unfortunately it is easier to deal with quantities than qualities, and it is almost impossible to comment on the latter faithfully here, but since our immediate purpose is to draw attention to some aspects of the matter which seem to be of special significance from the point of view of the possible processes of evolution in flowering plants, something can be achieved by considering the main types of families on the basis of the kind of generic constitution they show. On the three principles that species are in theory of equivalent value; that genera mark a considerably larger measure of differentiation than do species; and that the relative size of genera can be expressed by their species numbers, we can to some degree translate taxonomy into qualitative terms.

Types of families in the Flowering Plants

Type 1. Proceeding, as is more convenient, from small to large, the first type is the monotypic already mentioned, in which the family consists of only a single genus with a single species. There are some 50 of these, which means that this number of *species* have been adjudged so isolated from any others that they deserve family rank. Most of these strictly monotypic families are rather unfamiliar, but among them are Adoxaceae, Bixaceae, Cephalotaceae and Hippuridaceae.

Type 2. A second type of family is monogeneric but polyspecific, each consisting of a single genus with more than one species. These represent *genera* so isolated structurally that they are considered to merit family rank. These might be expected to outnumber the monotypes but it is nevertheless rather surprising to find that there are more than 70, or more than a sixth of the total. It is the numbers of their species which is of most interest here, and as again might be expected, though the point is not without evolutionary significance, the majority are quite small.

About 20 have only 2 species each, and of these the Byblidaceae, Roridulaceae and Punicaceae may be instanced. As many more have less than half a dozen species. Rather more than twenty have from six to thirty-five species and only nine have more than the latter number. These nine larger families are mostly familiar and all deserve mention. In the Cannaceae, which resemble the Gingers in many ways, there are some 50 species; the Casuarinaceae with 55 species, form one of the most peculiar families; the Nepenthaceae, 75 species, have very unusual vegetative characters; as also, in some degree, have the Musaceae, containing the 80 or so species of banana; the Grossulariaceae have 140 species, in the genus *Ribes*; the Cuscutaceae, also with about 140 species, are parasitic plants; the Potamogetonaceae, 160 species, afford the only notable instance of wide vegetative elaboration in freshwater habitats; the Sauraujaceae, with about 250 species, is a family of somewhat doubtful qualitative value; and the family Symplocaceae which, with 500 species, is by far the biggest, is also weak, and perhaps belongs more properly to the Rosaceae. In brief, of the nine larger monogeneric families, the Casuarinaceae, Nepenthaceae, Musaceae, and perhaps the Cuscutaceae, are the only ones which are really isolated groups in the highest sense of the word.

Type 3. A third type is the bigeneric family, which in terms of the stellar analogy may be likened to a double star. There are two kinds of these, one in which the two genera are about the same size in species number (or at least the disparity is not extreme having regard to the total numbers) and the other in which disparity between the genera is so great that their relationship can best be expressed as that of "planet and satellite". The former number about twenty but in only three of them, Betulaceae (80 spp., 40 spp.), Velloziaceae (70,45) and Schisandraceae (19,16) are the species in any sense numerous. On the other hand there are two families, Trichopodaceae and Huacaceae, in which each genus is monotypic. Other disparities are seen in Melianthaceae (30,6); Nolanaceae (30,8); Cabombaceae (7,1) and Hippocastanaceae (17,2). Families of this type fall into two according to the qualitative differences between the genera, some comprising relatively unlike genera, as in Betulaceae, Cannabinaceae, Elatinaceae and Calycanthaceae, and others genera which are more alike, as in the Hippocastanaceae. This last family leads to the second or satellite subtype and might even claim a place there, for the criterion is not only one of size disparity but also of intergeneric relation, but there are seven other families which are even more typical, namely Xyridaceae and Erythroxylaceae, each with 200 species in the larger genus; the Aceraceae, in which the number is 125; and the Tropaeolaceae (80), Clethraceae (55), Taccaceae (30), Stackhousiaceae (25), and Moringaceae (20). In almost all these the small genera are monotypic.

Type 4. This again is a satellite type and comprises three families, each containing one relatively very large genus and two or three monotypic genera. They are Balsaminaceae, in which the species numbers in the genera are 550:1:1; Plantaginaceae, in which they are 275:1:1; and Droseraceae, in which they are 90:1:1:1.

Type 5. The families in this type are also of the satellite kind, but have two larger genera, so that they are, as it were, double stars with satellites. These are, with the specific numbers of the genera:

Piperaceae 1400:800:3:1:1   Hippocrateaceae 170:100:2:1:1:1
Salicaceae 600:100:1:1   Alstroemeriaceae 100:55:1:1
Pandanaceae 300:150:1   Sabiaceae 100:30:1:1

Type 6. In the families of this type there is a relative paucity of both genera and species, though there are more than two of the former. It is a somewhat miscellaneous type, containing something like 70 families in all, and several subtypes can be recognized. In one, all the genera are monotypic, but this is rare and apparently only pertains in Philydraceae (4 genera), Achariaceae (4) and Dioncophyllaceae (3). In another subtype there are one or two larger genera while the rest are monotypic, and here there are about ten families, including Cyrillaceae, Basellaceae, Butomaceae, Diapensiaceae and Philesiaceae. In a third the genera are more equal in size, though none is large, and among these are Calyceraceae, Monotropaceae, Bruniaceae, Chlaenaceae and Empetraceae. A fourth subtype contains families with a considerably wider range of generic values and here there are upwards of twenty, exemplified by Juglandaceae, Alismataceae, Corylaceae, Resedaceae, Marcgraviaceae and Cyclanthaceae.

This sixth type in particular makes it clear that the proportion between the number of genera and the number of species in families is a matter of considerable interest from the evolutionary point of view, and this numerical relationship is best expressed as the species/genus ratio or S/G, obtained by dividing the number of genera into the number of species. Comments on the value of this expression in larger families will be made later but with regard to those already mentioned it may be pointed out that the lowest ratio, unity, is found where all the genera are monotypic, and that some of the families in the tabulation above have the highest of all ratios. In the Piperaceae it is no less than 441, far higher than in any other family, and the figures for Salicaceae, 173, and Pandanaceae, 150, are greater than in any family outside this type except the Begoniaceae.

One or two of the families so far mentioned, notably the Piperaceae, are large, but most of them are small, and as we move on to larger groups we must be careful not to put too much stress on the feature of size alone, for this, as estimated by the numbers of genera and species, reflects considerations of varying relevancy. It is also subject to a certain amount of variation at will, so that by changing taxonomic conceptions almost any result may be obtained, as witness the splitting up of *Mesembryanthemum* or the suggestion that all Crucifers really belong to one genus. At the same time it does appear that size and type of constitution go together to some extent, so that some features are shown better in small families and some better in large. For instance, the more numerous the genera of a family are the more complete is likely to be the series of species numbers in the genera, while conversely smaller families may or may not include one relatively large genus. It is particularly noteworthy evolutionarily that, with few exceptions, the greater the number of species in a family the greater also the number of genera. There is thus some justification for considering the size of families, using all due caution, as a relevant factor in grouping them, and the next three types of family illustrate this, because they are specially well represented among the medium sized families, those with less than one thousand species, which have next to be considered.

Type 7. This contains families in which no genus stands out very prominently in size and in which, consequently, the S/G ratio is low. There are more than fifty families here and though they form a fairly complete series they can be treated as being of two kinds.

In one of these each family has a single large genus which in many ways typifies the family, and here Dilleniaceae (*Dillenia*), Plumbaginaceae (*Statice*), Campanulaceae (*Campanula*), Lobeliaceae (*Lobelia*), Oleaceae (*Jasminum*), Onagraceae (*Epilobium*) and Tiliaceae (*Grewia*) are representative. The S/G of these families varies a good deal but on both sides of the average for the Flowering Plants as a whole, which is about 18. In the families of the other kind there is no outstanding genus so that the S/G tends to be lower, and is sometimes very low. Notable examples of this are Icacinaceae (S/G 4), Podostemaceae (4), Pedaliaceae (4), Hamamelidaceae (4·5), Haemodoraceae (5), Hydrocharitaceae (5·5), Phytolaccaceae (6), Simaroubaceae (6), Menispermaceae (6·5), Zygophyllaceae (8·5) and Monimiaceae (9). How far such low figures reflects something which distorts the impact of normal taxonomic methods on these families is an interesting speculation. The highest figures among these families are in Burseraceae (36) and Dipsacaceae (32), the latter being very unusual in having no genus with fewer than five species.

Type 8. In this type each family is dominated by one genus which, using an arbitrary measure, has at least three times as many species as the next in size. There are about forty of these among medium-sized families and with scarcely an exception the numbers of genera are under thirty. The most striking of these genus-dominated families relate closely to the satellite kinds, and are, with their corresponding figures:

Begoniaceae (*Begonia*)	800:10:3:2:1
Dioscoreaceae (*Dioscorea*)	500:20:4:3:2:1
Oxalidaceae (*Oxalis*)	500:51:15:7:3:2
Aquifoliaceae (*Ilex*)	400:11:3:3:1
Smilacaceae (*Smilax*)	300:7:6:2
Aristolochiaceae (*Aristolochia*)	300:28:12:8:6:1:1:1

Others scarcely less notable are Aizoaceae, Pittosporaceae, Santalaceae, Polygalaceae, Fagaceae, Orobanchaceae and Lentibulariaceae. In this type the S/G ranges from about 9 in Restionaceae and Cunoniaceae, which means that these families have long tails of small genera, to 66 in Vitaceae, 71 in Aizoaceae, 75 in Polygalaceae, 79 in Smilacaccac, 84 in Λquifoliaccae, 88 in Dioscoreaccae, 97 in Oxalidaceae, 125 in Fagaceae and 163 in Begoniaceae.

Type 9. This is a small group of families very similar to the last except that there are two dominant genera in each. Among the eight of these are:

> Caprifoliaceae (*Lonicera* and *Viburnum*)
> Combretaceae (*Combretum* and *Terminalia*)
> Ebenaceae (*Diospyros* and *Maba*)
> Juncaceae (*Juncus* and *Luzula*)
> Eriocaulaceae (*Eriocaulon* and *Paepalanthus*).

The S/G here is above average but not exceptionally so, the highest being 79 in Ebenaceae.

Larger families. There remain thirty families with between one thousand and two thousand species each, which we may call large, and two dozen with more than two thousand species each, which we may call the great families.

In the large families the main feature to be noted is the repetition, though in a modified form which may be attributed to the greater size of the families, of the types just discussed. Thus, although only four of these families are truly genus-dominated families, namely Araliaceae (*Schefflera*), Gentianaceae (*Gentiana*), Primulaceae (*Primula*) and Moraceae (*Ficus*), there are another half dozen or so in each of which there is one notably outstanding genus, among these being Gesneriaceae (*Cyrtandra*), Loranthaceae (*Loranthus*) and Myrsinaceae (*Ardisia*). There are also two which approach the two-dominant type, Caesalpiniaceae with *Cassia* and *Bauhinia*, and Urticaceae with *Pilea* and *Elatostema*. Most, however, show no markedly dominant genus, taking into account the size of the family, and this is particularly true of Cucurbitaceae and, less so, of Sapindaceae, Sapotaceae and Verbenaceae. The S/G in these thirty families varies, compared with earlier quotations, within comparatively narrow limits, from under 10 in Cucurbitaceae to 44 in Primulaceae, but there are few above 25.

The twenty-four great families of the Flowering Plants are so vast a subject in themselves that only one or two of their most relevant constitutional features can be referred to here, and this but briefly. First, it is clear from a survey of these families that they are not essentially different in type from those above. True, they cannot illustrate the points more peculiar to very small families, but otherwise parallels can readily be detected. Thus, the Piperaceae (already mentioned) is perhaps the most remarkable of star and satellite families: the Solanaceae is, in virtue of *Solanum*, strikingly genus-dominated, while in several others, notably Euphorbiaceae (*Euphorbia*) and Araceae (*Anthurium*) there is at least the sub-dominance of one genus; the Ericaceae strictly and the Mimosaceae virtually are double-genus-dominant families, the first by *Rhododendron* and *Erica* and the second by *Acacia* and *Mimosa*. On the other hand a greater proportion of these families are not dominated by any one genus to the same degree and in some of these, as in Asclepiadaceae, Umbelliferae, Cruciferae and Apocynaceae, the evenness of gradation in generic size is remarkable, even allowing for the large numbers of genera concerned. In the same way the S/G ratios are not notable. Two-thirds of them fall within 10 and 25; three are below this range, with 7·5 as the extreme, in Cruciferae; and five are above it, four of them being below 50 and the other being the exceptional 441 in Piperaceae.

The second point in which the great families are especially valuable exemplars, is one inherent throughout the Flowering Plants but which has deliberately been left till now in order to avoid confusion. This is the way in which they illustrate, not only some of the more strictly numerical aspects of differences in family constitution, but also much more important, though much more perplexing, qualitative differences. When these twenty-four families are reviewed carefully, if but rapidly, it is at once apparent that they differ considerably in certain respects not expressible merely in figures. In general these great families are sufficiently familiar to make this a matter of common knowledge, and we may proceed at once to a more direct statement of the facts. These families show, especially, three sorts of qualitative make-up which are of great interest in evolution. First, there is what may be properly called the specialized kind of family, in which the characteristic structural theme is peculiar and not found, at least in similar expression, elsewhere, and it is remarkable that the first three of the great families, the Compositae, with their peculiar pseudanthial organization; the Orchidaceae, the only polliniai family of monocotyledons; and the Papilionaceae, with unique zygomorphy, are of this kind, to which may be added the fifth family, Gramineae. Besides these there are, further down the list at least two other specialized families, the Asclepiadaceae, the only pollinial family of dicotyledons (see Chapters 7 and 8), and the Araceae, another unique pseudanthial type (see Chapter 9). Specialized families tend, by the nature of the case, to give an impression of homogeneity.

Second, there are strongly patterned or homogeneous families, in which specialization consists not so much of extreme exploitation of a special theme as of close adherence to a particular but not outstanding plan. To the taxonomist the result is often much the same, particularly in the difficulty of breaking the family down into genera, but there is a real difference between the two sorts, and those of which we speak are essentially homogeneous rather than specialized. This becomes plain enough when the families are quoted, being, in descending order of size, Labiatae, Cyperaceae, Melastomataceae, Acanthaceae, Umbelliferae, Cruciferae, Piperaceae, Mimosaceae and Palmae. Our point may be reinforced by saying that in these families perhaps less than in any others is there any practical confusion or doubt as to what species belong to them and what do not. Perhaps one other family, Myrtaceae, should be added to the list.

Third, half a dozen families are in effect the converse of those just listed and may on this account be conveniently called heterogeneous families, since in each the content is most remarkable for its lack of constancy. These are the Rubiaceae, Euphorbiaceae, Scrophulariaceae, Solanaceae, Apocynaceae and Rosaceae, and here again we may make our point more vividly by inviting the reader to say which is the most typical member of each of these families. In some of these there is a certain amount of basic general characterization, as in the interpetiolar stipules in Rubiaceae, or in the ovaries of Solanaceae and Euphorbiaceae, but this is masked by the overall heterogeneity of form. Yet despite all this, few taxonomists will deny or doubt that most of these families (the Euphorbiaceae chiefly excepted) are satisfactory natural groups in the best sense of that much-abused phrase, and that they have so unerringly detected and delineated them is not only much to their credit but clear evidence that in spite of all their artificialities there is fundamental truth in most modern classifications. There remain two great families that have not yet been placed, the Liliaceae and the Ericaceae, and the difficulty with these is that they are partly homogeneous and partly heterogeneous. On the whole it may be claimed that they are more the latter than the former, but it is of greater significance that these two have long been flies in the ointment of taxonomists, who are conscious that, as total unities, they are unsatisfactory.

What has been said about the great families applies in some degree throughout the Flowering Plants, and even among the smallest families we can detect the more specialized and less specialized homogeneous and the heterogeneous, and it is on this note that our preliminary consideration of the families may best end for it is expressive of something of first-rate importance in the understanding of their evolution. Whatever the processes by which this may have come about, they have, by and large, repeatedly produced one or other of three main sorts of major groupings. In one the accent has been on the exploitation of a particular and often unique specialization. In another the stress has been on the elaboration of a comparatively restricted theme. In yet another it has been on the production of difference rather than similarity. If it has done no more this review of flowering plant families has demonstrated how various may be the results of evolution, and how various, one may venture to hazard also, may have been the evolutionary processes themselves.

NOTE ON LITERATURE

The most complete and fully illustrated source of information about the Flowering Plants are the first, and partially published second, editions of

Engler, A. and Prantl, K. *et al.* (ed.): *Die natürlichen Pflanzenfamilien*, Leipzig, 1889–1915 and 1924– ,

and the numerous volumes of

Engler, A. (ed.): *Das Pflanzenreich*, Leipzig, 1900– .

The most up-to-date, though unillustrated, work on the group as a whole is

Lemée, A.: *Dictionnaire déscriptif et synonymique des genres de plantes phanérogames*. Brest, 1929–1943– .

Supplementing the three above chiefly on account of its fine illustrations is the much older

Baillon, H. E.: *Histoire des plantes*, 12 vols., Paris, 1867–94, and the translation of its first eight volumes by M. M. Hartog under the title *Natural History of Plants*, London, 1871–88.

For coloured illustrations of individual species there is the incomparable
Curtis, W. *et al.*: *Botanical Magazine*, London, 1793–   .

Valuable, shorter and well illustrated accounts of the Flowering Plants are
Rendle, A. B. : *The Classification of Flowering Plants*, 2 vols., Cambridge,
   1925 and 1930,
Hutchinson, J.: *The Families of Flowering Plants*, 2 vols., London, 1926
   and 1934,
and, among the most recent,
Lawrence, G. H. M.: *Taxonomy of Vascular Plants*, New York, 1951.
Finally there is
Willis, J. C.: *A Dictionary of the Flowering Plants and Ferns*, 6th edition,
   Cambridge, 1931,
which contains more concentrated information than any other work of its
size and which is an indispensable handy reference book.

CHAPTER FIVE

A REVIEW OF THE FLOWERING PLANTS

II. MONOCOTYLEDONS

SINCE no two individual plants are precisely alike it might be expected that, in any attempt to analyse a group of plants into its biological kinds by the methods suggested in the last chapter, there would arise the difficulty of deciding, other than arbitrarily, where to stop the process. This is not, however, a great problem because the discontinuity in nature is such as to present successive crucial stages in any progressive analysis of this sort. That is to say a group of organisms can, on any given set of criteria, be divided up only to a certain point, beyond which it becomes necessary to employ new values and new criteria, and it is this which imparts to the familiar taxonomic framework of families, genera and species, much of such reality as it possesses. Not only so, but the further an analysis is carried the more numerous are its products, so that there must always be a practical point at which, according to circumstances, the advantage of any further dissection is out-weighed by the increase in complexity. Finally, it will be clear that what we have referred to as the crucial stages in analysis will correspond fairly closely with the use of the main taxonomic categories.

All this becomes readily apparent when the Monocotyledons as a whole are examined in the way outlined earlier, for it is soon found that the first crucial stage reveals some thirty different biological sorts, or groups, of these plants, and that of these the majority are almost or quite congeries of individuals which the taxonomists have recognized as families. Had the aims of taxonomy and of the present analysis been exactly the same then there might well have been complete coincidence between the families and these biological groups, but as it is the number of the former is much the larger because there are used in classification additional diagnostic characters which are not of strict biological significance, as is familiar enough, for instance, in the liliiflorous monocotyledons. It cannot, of course, be claimed that there is anything immutable about this number thirty, and indeed other personal estimates might well make the total rather more or less but it does seem that the recognition of about this number of sorts of plants reveals the essential pattern of the Monocotyledons as a whole with as great a combination of truth and simplicity as can reasonably be expected.

There is only one more point calling for clarification before we go on to the presentation of these constituent groups which is the main purpose of this chapter. It was explained earlier that the task here is really to point out those sorts of plants which make appreciable and biologically distinctive contributions to the biota and its environment as a whole, and the inclusion here of the word *appreciable* means that the factor of prevalence must be taken into account. The conception here is quite straightforward. If a kind of plant is ubiquitous and plentiful (e.g. the Grasses) it will make a large contribution of its own to the vegetation, even if its distinctiveness from all others (e.g. the Sedges) is not, in every respect, very great. If, on the other hand, a kind is sparse in quantity and local in distribution, its contribution will be small however peculiar it may itself be, and it will not for this reason, merit separate recognition.

Partly because in many cases no taxonomic name exactly fits them, and partly to emphasize that there is no necessary coincidence with any particular families or other taxa, each of the groups in the following presentation is given a generalized English or anglicized name, spelt with a capital letter. Where there are sufficiently familiar and unambiguous English names these have been used; in some cases straightforward taxonomic names have merely been anglicized; in some the suffix -ads has been applied to the name of the most convenient, or only contributory, taxon; in a few, where nothing else seems to meet the case, a brief descriptive phrase is used. The purpose of these names is purely utilitarian, that of allowing the groups concerned to be referred to as cogently as possible.

In view of the imperative need to conserve space no attempt has been made to describe the characteristics of each group in detail, or even to diagnose it at length, but the taxonomic content of each is stated and this should sufficiently indicate what larger taxa are included. Further information about these should be sought in text-books of formal taxonomy such as those listed at the end of Chapter 4.

The Chief Groups of Monocotyledons

1. *The Palms* (*fig.* 1).

One of the best-known and most distinctive groups and the most considerable and striking example of woody growth in monocotyledons. The often bulky columnar stems, huge leaves, and frequently enormous multiflorous inflorescences give the plants an ecological and biological potential which is unique in that series, except, perhaps, for some of the larger Bamboos. This pantropical group comprises the erect members of the families Palmae and Cyclanthaceae, totalling in all about 1,650 species. Some of the smallest and least woody members approach the state of some of the Aroids (group 14).

2. *The Rattans* (*fig.* 2).

A satellite group closely connecting with the Palms proper and containing the climbing members of the same two families. Their growth form is peculiar and ecologically important in the tropics, where all their 375 species are found.

3. *The Screw-pines* (*fig.* 3).

Here the main visual features are the woody habit (often branched), numerous simple leaves in a spiral arrangement, unisexual flowers and multicarpellary fruits. This group comprises the family Pandanaceae of the Old World tropics, with about 450 species. Some species of the genus *Freycinetia* are actually climbers, repeating the kind of relation that exists between the Palms and Rattans, but these are not significant enough to be separated.

4. *The Bananas* (*fig.* 4).

These are the giant leaf-stem herbs so prominent in many kinds of tropical vegetation. The genus *Musa*, containing the true bananas, forms the core of the tropical and mostly Old World group, but *Stretlitzia* and *Heliconia* must also be included. *Ravenala*, though anomalous in having a columnar woody stem, is best added on the criterion of the leaves. There are about 140 species in this group.

5. The Velloziads *(fig. 5)*.

A small group containing the American and African tropical family Velloziaceae with about 115 species. These plants have often branched, usually slender, woody stems of somewhat unusual structure; innumerable very narrow leaves, and generally solitary flowers.

6. The Tree-lilies *(fig. 6)*.

This seems to be the most satisfactory name to give to those plants which comprise the remainder of the notably woody-stemmed monocotyledons, and which nearly all fall within the four families Agavaceae, Xanthorrhoeaceae, Bromeliaceae and Liliaceae. They are essentially tropical plants and it is difficult to estimate their numbers closely because there is such variation in size, but about 375 species seem to merit inclusion here. *Cleistoyucca arborescens*, with deliquescent branching, is one extreme condition and the huge columnar species of *Puya* another. Some species of *Cohnia* and *Dracaena* can only be described as miniatures of this sort.

7. The Grasses *(fig. 7)*.

This most familiar group, comprising the more herbaceous species, about 7,000 in number, of the family Gramineae, needs no further description. Their combination of vegetative and floral characters strongly isolates them (except from the Bamboos), but when the latter features are unusually simple, as in *Anomochloa marantoidea*, there is a distinct resemblance to some of the Gingers (group 13).

8. The Bamboos *(fig. 8)*.

A satellite group of the last (as the Rattans are of the Palms) comprising the 500 or so species of the woody tropical subfamily Bambuseae of the Gramineae. Here, however, the satellite has the larger and bulkier growth-form.

9. *The Sedges* (*fig.* 9).

A familiar group comprising the cosmopolitan family Cyperaceae with more than 4,000 species. The Sedges show less specialization than do the Grasses, of which they are, in certain respects, a more aquatic ecological counterpart.

10. *The Pipeworts* (*fig.* 10).

This group, which is entirely made up of the widely distributed, but mainly tropical and helophytic family Eriocaulaceae of about 600 species, is of special interest because of its pseudanthial capitulate inflorescences (see Chapter 9).

11. *The Rushes* (*fig.* 11).

This group comprises not only the plants which habitually go by this name and which make the family Juncaceae, but also the southern family Restionaceae, in which the separation of the sexes in the flowers does not detract from a close resemblance in stature and design. The small families Centrolepidaceae and Thurniaceae are also best included here, and the sum total is about 650 species.

12. *The Orchids* (*fig.* 12).

This is by far the largest really coherent group of monocotyledons in species numbers, and comprises the Orchidaceae, which, with 17,500 species, is the second largest family of flowering plants. It is interesting to note that while in floral structure and biology the orchids are quite isolated, they include members which, superficially at least, tend to link up with such other groups as the Gingers (next), the Saprophytic monocotyledons (group 21), and even some others of groups 18 to 29.

13. *The Gingers* (*fig.* 13).

This is the best general name for the very distinctive pantropical group composed of the families Zingiberaceae, Marantaceae and Cannaceae, to which the small Apostasiaceae and Lowiaceae may be added, and which amounts in all to some 1,500 species. The chief features are the characteristic, often rhizomatous habit; frequently broad-bladed foliage: and strongly zygomorphic floral specialization—but some of the members are strikingly anomalous.

14. *The Aroids (fig. 14).*

This, which with one or two minor exceptions, comprises the widespread but predominantly tropical family Araceae, with some 2,000 species in all, is the most striking and important pseudanthial group in the Monocotyledons, and, if only for this reason, occupies a a very isolated position. The leaves are also often very characteristic.

15. *The Large-flowered Aquatic Monocotyledons (fig. 15).*

This is a rather heterogeneous group comprising those helophytic and more fully aquatic monocotyledons which generally have relatively large and conspicuous heterochlamydeous regular flowers. In floral design they are near the Spiderworts and in floral size they link with the Small-flowered Aquatic Monocotyledons next to be mentioned. They total about 450 species, making up some or all of each of the families Alismataceae, Butomaceae, Hydrocharitaceae, Mayacaceae, Philydraceae, Pontederiaceae, Rapateaceae and Xyridaceae, and they are found in all parts of the world.

16. *The Small-flowered Aquatic Monocotyledons (fig. 16).*

This is also a heterogeneous world-wide group, including three rather distinctive components, but no one of these can really be separated as a group of its own. The first and second of them are, respectively, the more or less submerged freshwater aquatics and the submerged marine angiosperms belonging to the families Aponogetonaceae, Naiadaceae, Posidoniaceae, Potamogetonaceae, Ruppiaceae, Zannichelliaceae and Zosteraceae. The third contains the rather varied helophytes and larger aquatics of the families Heterostylaceae, Juncaginaceae, Scheuchzeriaceae, Sparganiaceae and Typhaceae as well as one or two aroids. The Juncaginaceae have much in common biologically with the Plantagos among the dicotyledons; the Typhaceae with such grasses as *Phragmites*, while *Sparganium* and the aroids link vegetatively with group 15. The group has about 335 species, of which about half belong to *Potamogeton*.

17. *The Duckweeds (fig. 17).*

The only "thalloid" monocotyledons. Tiny free-floating aquatics with very simple, rather infrequent flowers, comprising the cosmopolitan family Lemnaceae, most usually reckoned to contain only about a dozen species.

18. *The Bromeliads* (*fig.* 18).

The typical member of this group is a tropical plant, often epiphytic, large-herbaceous in stature, and of a characteristic facies, having a rosette of comparatively few, stout or subfleshy, often strongly toothed persistent leaves with sheathing bases, and a multiflorous inflorescence of generally heterochlamydeous flowers. Its main component is provided by the medium-sized members of the almost entirely American family Bromeliaceae, but to this must be added a large ingredient from the Liliaceae, notably many of the genus *Aloe:* another from the Agavaceae, especially from *Agave* and *Sansevieria:* some from the Xanthorrhoeaceae; and, on general habit, a few of the Commelinaceae and Iridaceae. The smaller Bromeliaceae also include a number of rather anomalous plants, notably in *Tillandsia* and *Nidularium* but these can hardly be treated separately. The group as a whole is pantropical and contains about 2,000 species.

19. *The Spiderworts* (*fig.* 19).

These are subtly different from almost all other monocotyledons, though they have few positively distinguishing features. The group contains about 500 species and comprises the more generalized majority of the pantropical family Commelinaceae, in which the most prominent characteristics are the vegetative design, the boat-shaped bracts below the inflorescence, and the strongly heterochlamydeous flat flowers.

20. *The Taccads* (*fig.* 20).

A small and unfamiliar group, comprising the 30 or so species of the pantropical family Taccaceae, which, on account of their characteristic combination of leaf and flower values, can neither well be ignored nor included in any other group. The foliage has some link with the Aroids but the flowers have more in common with some of the groups 22 to 29.

21. *The Saprophytic Monocotyledons* (*fig.* 21).

This is admittedly one of the more artificial groups and is necessary to account for the various non-green saprophytic monocotyledons of the five families Burmanniaceae (with certain exceptions), Corsiaceae, Petrosaviaceae, Thismiaceae and Triuridaceae. Of the 170 or so species, in all widespread, some show little peculiarity of structure; others resemble certain Orchids; and some are extremely bizarre.

22. *The Asphodels* (*fig.* 22).

These include the plants often thought of as the most typical monocotyledons, namely geophytes with a basal rosette of usually non-persistent narrow leaves and an emergent racemose inflorescence of numerous homochlamydeous flowers. The size of this group is not easy to estimate, partly because it merges so completely with others and partly because it extends over so many taxa. It is world wide in distribution and seems to contain about 2,400 species, nearly all within the wider conceptions of the families Liliaceae, Amaryllidaceae and Iridaceae. Much as do the Bromeliads, the Asphodels include some rather anomalous plants, such as *Massonia* and *Daubenya*.

23. *The Amaryllids* (*fig.* 23).

This is really a twin of the last and is so like it in many ways that it is hard logically to justify its separation, the verbal difference being little more than that the inflorescences are umbellate instead of racemose. This difference is, however, clear cut and often accompanied by intensifying minor features which makes the averages of the two rather unlike. This group also is world-wide but has only about 1,200 species. Its taxonomic distribution is also much as the last and likewise it merges with several other groups.

24. *The Tulips* (*fig.* 24).

This comprises the otherwise Asphodel- or Amaryllid-like plants which have, either invariably or habitually, only a single flower. Defined thus it seems to contain about 500 species, again mainly in the same three families, but it merges very completely with one or two other groups especially the Amaryllids. It is widely distributed but notably associated with northern extra-tropical regions.

25. *The Irises* (*fig.* 25).

This group, with its characteristic kind of heterochlamydy is almost restricted to the genus *Iris*, but a few related others bring the species total to about 280 with a rather incomplete world-wide distribution. Structurally it merges most completely with the Asphodels, and ecologically with the Large-flowered Aquatic Monocotyledons.

26. The Philesiads (fig. 26).

This rather small but widespread group of about 250 species comprises the herbaceous liliiflorous monocotyledons in which aerial stems bear scattered leaves, in the axils of which the flowers, generally large, appear singly or in small groups. The most fully characteristic are probably the genus *Philesia* and the erect species of *Luzuriaga*, but there are more familiar members in *Polygonatum* and parts of the genus *Lilium*.

27. The Phylloclade Monocotyledons (fig. 27).

Here are included two rather distinctive-looking sorts of plants, the butcher's-brooms of the Ruscaceae, and the genus *Asparagus*, but both share the same unlikeness to other monocotyledons in the nature of their photosynthetic surfaces. The species number about 150, of which the Ruscaceae contribute only a few, and they have a wide distribution in the Old World.

28. The Large-flowered Climbing Monocotyledons (fig. 28).

These comprise the larger-flowered, more or less homochlamydeous climbing liliiflorous plants, namely the genera *Bomarea*, *Stemona*, *Gloriosa*, *Behnia*, *Lapageria* and some species of *Luzuriaga*, the first containing a very great majority of the 150 widely distributed species. Some of the more erect species of *Bomarea* should more strictly perhaps fall under the Philesiads.

29. The Small-flowered Climbing Monocotyledons (fig. 29).

This is a much larger group than the last, having 850 species, nearly all in *Dioscorea* and other genera of the Dioscoreaceae and in *Smilax* and other genera of the Smilacaceae. With few exceptions they have unisexual and dioecious flowers and are best described as pantropical.

In producing the foregoing analysis of fewer than thirty components the monocotyledons have naturally been painted with a broad brush, but it is a fact of no small interest from the point of view of their evolution, that, allowing for the several anomalous kinds of plants mentioned under some of the groups, there are very few monocotyledons indeed which do not find a place in one or other of these twenty-nine groups. Of these, the genera *Convallaria*, *Maianthemum*, *Lourya*, *Aspidistra*, *Curculigo* and *Molineria*: the peculiar large herbaceous genera *Joinvillea* and *Susum* in Flagellariaceae: the anomalous bulbous climbing plants in *Bowiea* and *Schizobasis* in Liliaceae: and the three various genera *Calectasia* (Liliaceae), *Trichopus* (Trichopodaceae), and *Lilaea* (Heterostylaceae) are perhaps the most noteworthy.

Conspectus

As the most practical way of summarizing the foregoing presentation of the monocotyledons the rest of this chapter is devoted to a consideration of two particular aspects of it—the arrangement of the groups, and their more pictorial representation.

It will be apparent enough that the order in which the groups have been set out brings together those which have certain outstanding features in common. Groups 1 to 6, for instance, include nearly all the exceptionally large and robust monocotyledons, a fact which may be expressed otherwise by saying that in them the predominant theme is one of size and woodiness. Rather similarly, though conversely, groups 15 to 17 comprise those monocotyledons which are the most completely aquatic, and with some of which there is associated extreme lack of size. Falling, in some respects, rather between these two are groups 7 to 11, in which the most characteristic feature is simplicity, of both size and form, in the flowers which, in these less-aquatic plants, are either themselves small and scarious, or developed among glumes of this description. In contrast groups 12 to 14 comprise the three great sections of the monocotyledons which show marked and positive floral specialization, in two cases expressing itself as zygomorphy and in the third as pseudanthy. In short these four sets of groups illustrate as many examples of that general sort of structural or organizational theme which is often, though loosely, spoken of as an evolutionary tendency, though it begs the question of the direction of change. More accurately they show certain conditions that tend to be expressed largely or predominantly in what has presumably been the evolutionary elaboration of the present monocotyledons from their ancestors. Of course this is not the only way in which the groups might have been arranged, and the treatment here involves, and indeed rests upon, the acceptance of the opinion that these features are of more significance than others. But realizing how important in one way or another these are in determining the contribution of their possessors to the overall environment, this view appears not only justifiable but also difficult to dispute.

It is important to notice that, easy as this arrangement may be to make, its themes are not wholly and mutually exclusive. Thus, size is expressed also in the Bamboos, but these cannot on this account, be removed from the evolutionary proximity of the Grasses, while conversely, the Rattans can scarcely be regarded otherwise than as Palms without the dimensions more normal for that group. Similarly some of the more extreme aquatics have flowers similar in size and organization to those in groups 7 to 11, or to those of the Aroids. On the other hand, the overall floral design of the Orchids, Gingers and Aroids serve to make them the most isolated of all the groups.

Groups 18 to 29 are rather different from the rest. It will already have been appreciated that groups 1 to 17 comprise the plants which, if the monocotyledons were dismembered by a process of elimination, would first and most readily be subtractable, and a glance at groups 18 to 29 is enough to confirm that these form what we may usefully call a "harder core" of the monocotyledons, the value of this phrase lying in the way in which it stresses the increased degree of cohesion between the groups involved. The conception of a core is happy also because these groups comprise those monocotyledons which, on almost any diagrammatic representation, find a central position relative to the rest. These are indeed the monocotyledons in which are to be found, toned down towards a general mean, most of the more extreme expressions notable in the other groups. They are, for instance, in the general run neither very large nor very small, nor extreme aquatics, nor are their flowers either very simple or very specialized. In short, their characterization, in comparison with the rest, is negative rather than positive.

But this must not be taken to mean that all these last twelve groups are much alike, for this is far from being the case. The point is rather that they are relatively alike in terms of the character dimensions which distinguish the earlier groups, so that though they can be fairly easily subdivided, the criteria on which this is to be done are smaller than those used in connection with the others. In a word, groups 18 to 29 are individually less specialized and therefore less obviously to be separated from one another. At the same time they are by no means all equally alike or unlike, and they fall readily into two sets, in the first of which the groups (18 to 21) are on the whole more different from one another than are those (22 to 29) of the second. These last do indeed provide the final close-knit heart of the monocotyledons, yet nevertheless a glance at the names shows how various are the plants concerned. Nor do groups 18 to 29 appear to illustrate any very definite structural theme of their own, except that they include, particularly in the second set, most of the monocotyledons in which the perianth is homochlamydeous and petaloid, and most of those which have bulbs: but the former is something of a negative character from the point of view of differentiation, while the latter is widely paralleled elsewhere by other kinds of subterranean perennating organs. Perhaps the most particularized plants of these groups are the members of the genus *Asparagus*, with their peculiar photosynthetic parts. The Saprophytic monocotyledons also include some very bizarre plants but these are, on the whole, rare and inconspicuous.

We must now turn to the last item of this chapter, the question of the representation of the facts of our analytical survey in more graphic terms. It is presumably evident enough from what has been said that the framework of the review is simple, that is to say the number of groups is small and their separations sufficiently wide to make them readily expressible in diagrammatic form. Further, the remarks made about relative specialization point the form that such a diagram will take. Its central feature must clearly be the mass of groups 22 to 29, arranging themselves with the Asphodels and Amaryllids centrally, and expressed either to scale, according to their species content, or more conventionally by some other means. Immediately round this middle feature must be placed groups 18 to 21. Outside these again the remaining groups form another, and still more peripheral, zone. The resultant diagram, in which each group is shown by a *circle* of appropriate size, namely of a diameter which is the square root of its species number, is shown in fig. 30.

Interesting as this diagram is, it yet has at least one obvious limitation, namely that it cannot show every group in its approximately proper position in relation to every other. The groups in each of the two rings can, it is true, be placed in any desired circumferential order, but no arrangement of this sort can put all of them in their correct mutual positions, according to their total morphological resemblances and differences.

Much of what cannot be shown in a two-dimensional diagram of the sort described, can, however, be shown by means of a three-dimensional model, and it is perhaps the most noteworthy feature, from the evolutionary point of view, of this analysis of the monocotyledons, that it can be expressed with such comparative ease in this way. It would, of course, be absurd to claim that we know enough about the past to make such a survey or its model tell the whole truth, but the number of groups recognisable on the scale employed is so small, their separation is on the whole so decisive, and their mutual characterisation so simple in general outline, that it is not difficult to make a model in which each group is represented by a *sphere* of appropriate size (namely of a diameter roughly the cube root of the species number) which at least will show us a great deal about the monocotyledons, and which will help us to visualize both them and some aspects of their possible evolution in novel fashion. Moreover, as fig. 31 shows, this model can be reproduced in two dimensions without any part of it becoming wholly obscured.

One final point. Since it has been the purpose of this chapter, first, to analyse the monocotyledons and then to depict them in diagrammatic and model form, it is apparent enough that the model shown in fig. 31 is one of that great assembly of plants, but it is worth pointing out that if it were completely removed from its context here it might easily be identified, or even used, as a model of a constellation or star cluster in the heavens. Thus we see how an attempt to depict the monocotyledons, superficial as it doubtless must be, not only reveals many of their most interesting evolutionary features, but reinforces the value of the *stellar analogy* in problems of this kind.

FIG. I. A Palm: *Seaforthia (Ptychosperma) elegans*, very greatly reduced, after Le Maout and Decaisne

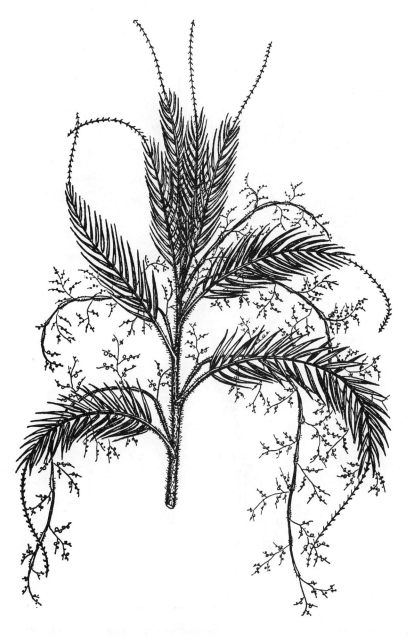

FIG. 2. A Rattan: *Calamus adspersus*, × ⅓, after *Pflanzenfam*.

FIG. 3. A Screw-pine: *Pandanus candelabrum*, very greatly reduced, after
Le Maout and Decaisne

FIG. 4. A Banana: *Musa* sp., $\times \frac{1}{20}$, after Brown

FIG. 5. A Velloziad: *Vellozia brevifolia*, slightly reduced, after *Pflanzenfam*.

FIG. 6. A Tree-lily: *Yucca gloriosa*, × ⅛, after *Pflanzenfam*.

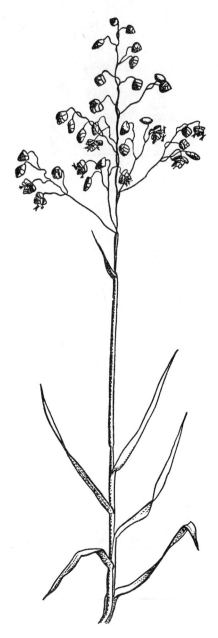

FIG. 7. A Grass: *Briza media*, × ¾

FIG. 8. A Bamboo: *Bambusa thouarsii*, very greatly reduced, after Le Maout and Decaisne

FIG. 9. A Sedge: *Carex* sp., × ⅓

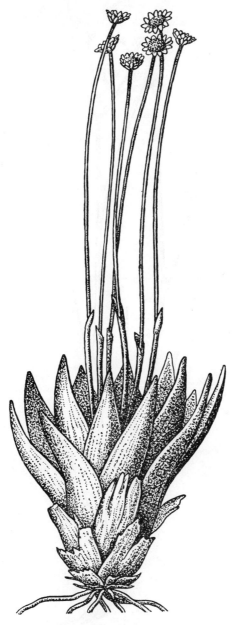

FIG. 10. A Pipewort: *Mesanthemum rutenbergianum*, $\times \frac{1}{2}$, after *Pflanzenfam*.

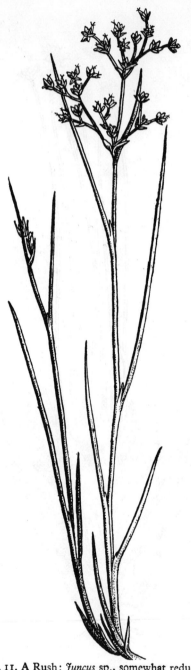

FIG. 11. A Rush: *Juncus* sp., somewhat reduced

FIG. 12. An Orchid: *Angraecum superbum*, × ⅛, after Bailey

FIG. 13. A Ginger: *Zingiber officinale*, × ⅓, partly after *Pflanzenfam.*

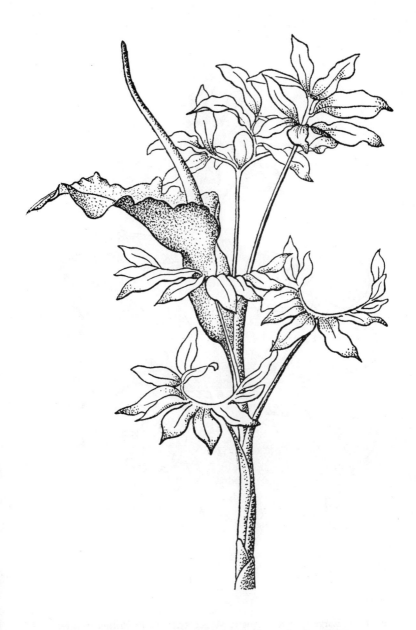

FIG. 14. An Aroid: *Dracunculus vulgaris*, × ¼, after Bailey

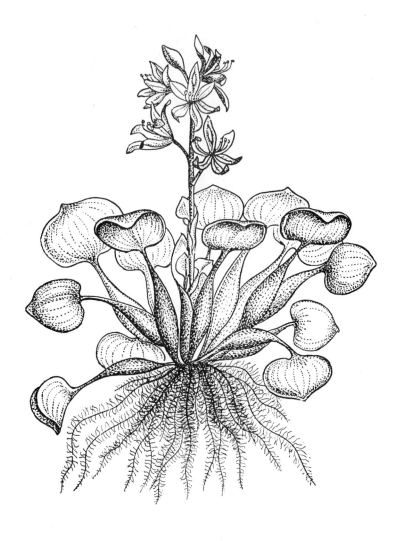

FIG. 15. A Large-flowered Aquatic Monocotyledon: *Eichhornia crassipes*, × ½, after *Pflanzenfam*.

FIG. 16. Freshwater and marine Small-flowered Aquatic Monocotyledons: *Potamogeton gramineus*, natural size, and *Zostera marina*, × ½, after *Pflanzenfam*.

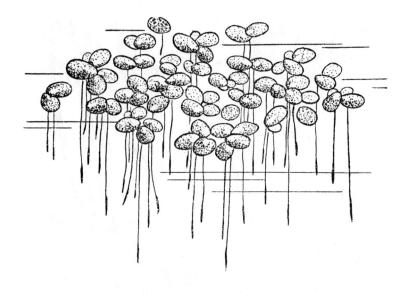

Fig. 17. A Duckweed: *Lemna minor,* × 2

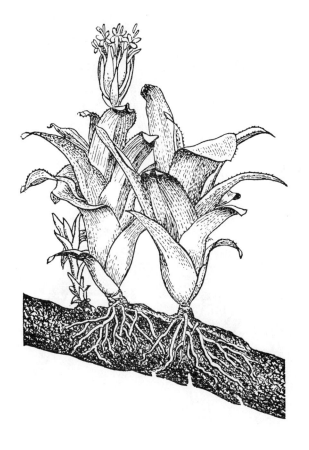

FIG. 18. A Bromeliad: *Billbergia* sp., $\times \frac{1}{7}$, after Brown

FIG. 19. A Spiderwort: *Tradescantia congesta*, natural size, after Maund

FIG. 20. A Taccad: *Tacca leontopetaloides*, $\times \frac{1}{7}$, after *Pflanzenfam.*

FIG. 21. A Saprophytic Monocotyledon, *Sciaphila maculata*, × 1½, after
Pflanzenfam.

FIG. 22. An Asphodel: *Asphodelus albus*, × ⅓, after *Pflanzenfam*.

FIG. 23. An Amaryllid: *Crinum powellii*, × ⅛, after Bailey

FIG. 24. A Tulip: *Tulipa* sp., $\times \frac{1}{2}$

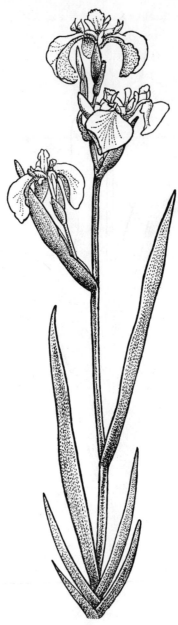

FIG. 25. An Iris: *Iris* sp., × ⅓

Fig. 26. A Philesiad: *Luzuriaga erecta*, natural size, after *Pflanzenfam*.

FIG. 27. A Phylloclade Monocotyledon: *Ruscus hypoglossum*, × 1½, after **Kerner** and Oliver

FIG. 28. A Large-flowered Climbing Monocotyledon: *Gloriosa virescens*, **natural** size, after *Pflanzenfam.*

Fig. 29. A Small-flowered Climbing Monocotyledon: *Smilax glauca*, × ¾, after Bailey

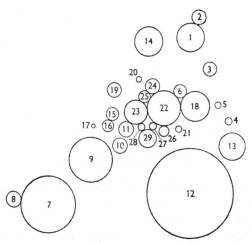

FIG. 30. A Diagrammatic Representation of the Monocotyledons (see p. 86)

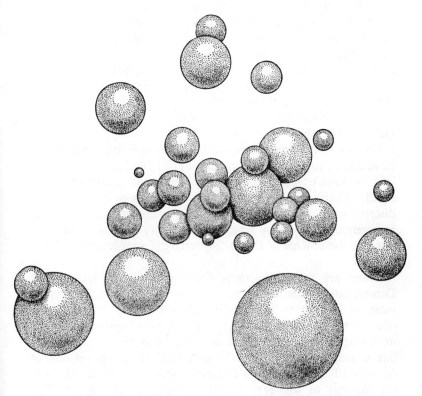

FIG. 31. A Drawing of a Stellar Model of the Monocotyledons (see p. 87)

CHAPTER SIX

A REVIEW OF THE FLOWERING PLANTS

III. DICOTYLEDONS

THE same sort of stellar diagram and model might be made for the Dicotyledons as was made for the Monocotyledons, but it is impracticable to attempt this here because the former are so much larger and more complex a series. Fundamentally, as Chapter 3 showed, the two are similar and their individuals are related to one another in much the same way, but the Dicotyledons, partly, no doubt, because they possess certain features conspicuously lacking in the Monocotyledons (such as open bundles and annular cambium; true leaf blades; and more varied heterochlamydy) comprise very many more taxa much more intricately related. In consequence it would scarcely be possible to make a model, or even a diagram, of them simple enough to allow adequate reproduction in two dimensions only. They must therefore be treated rather differently.

Because of their greater numbers and complexity it will clearly be necessary to intensify the process of simplification, and this can best be done by categorizing their constituent groups still more rigidly. Their complexity also makes it less easy to speak always of discrete units and more use must be made of the concept of representative types, by which the necessity of finding an exact place in the scheme of things for every different kind of plant is avoided.

There are many directions from which an analysis of the Dicotyledons on these lines might be approached but perhaps the most useful, bearing in mind the general purpose here, is by way of a short consideration of what is usually called specialization. In its biological sense this is not altogether easy to define precisely but it may be more generally described as the possession of features different from, and of a more peculiar sort than, those of the majority of whatever totality is under discussion, and this description has the advantage of drawing attention to two very important aspects of specialization, namely that it is not only particularization but also departure from a norm or average. Neither of these can exist alone, for there cannot be particularization without prior generalization nor an average without prior disparity, and so it is evident that we are likely to make good progress towards our goal of dissecting the Dicotyledons if we deal first with their specializations and then with the more generalized or average remainder.

The brief survey of families in Chapter 4 showed that most of the major groups of flowering plants belong to one or other of three types, which may be called the specialized, the strongly patterned homogeneous, and the heterogeneous. In the first the unifying character of each tends to be unique and therefore un-repeated; in the second unification rests chiefly on close adherence to a finely drawn, though not necessarily highly peculiar, plan; and in the third the unifying characters tend to be over-ridden and concealed by the diversity of the rest, so that there is no consistent visual similarity among the plants concerned. This simple classification must not, however, be pressed too far because the three types are neither wholly inclusive nor mutually exclusive. All specialized groups are, by their very nature, to some extent homogeneous, and conversely many homogeneous, and even heterogeneous, groups must, either by their very adherence to plan or by the peculiarity of some of their characters, be regarded as more or less specialized. For instance, the Labiatae were described as homogeneous, but they can equally claim a considerable measure of peculiarity, and in the same way the Scrophulariaceae, though they are markedly heterogeneous, show, in the zygomorphy of many of them, notable specialization. But such considerations must neither deter us from using these criteria, which have a considerable practical value, nor blind us to the possibility that they may possess a profound evolutionary significance.

It was also noted in Chapter 4 that the kind of family which is commonly regarded as specialized is, perhaps rather unexpectedly, conspicuous among the largest, and this merits emphasis here. Chapter 5 showed how many of the Monocotyledons are accounted for by specialized groups (and this quite irrespective of the question whether monocotyledony is itself a specialization), this being sufficiently illustrated by saying that the Orchids and Grasses, each of which have most unusual features, amount to more than 30 per cent. and 16 per cent. respectively of all Monocotyledon species, and, as will be seen, a similar state of affairs pertains in the Dicotyledons. True, no family here contains more than 10 per cent. of the whole series but it is noteworthy that the largest family, the Compositae, is always regarded as one of the most specialized, as is also the second largest family, the Papilionaceae.

But, as was said earlier, specialization combines two rather distinct conceptions, namely peculiarity of design and structural isolation. It is true that some of the more extreme conditions in a variable series or variation gamut may be, and indeed often are, described as specializations but the word is more strictly to be applied to those peculiar states which are not connected with less egregious types by intermediates. At any rate structural isolation is an important ingredient in, even if it is not a *sine qua non* of, specialization, and it is in this connection that there is seen to be one of the more subtle, but at the same time most profound differences between the Dicotyledons and the Monocotyledons. In the latter many groups, such as the Orchids, Aroids, Palms, Bananas and Grasses, are almost or totally unconnected by true intermediates, either with each other or with the rest of the Monocotyledons, which form as has been said a comparatively small proportion of the whole. That is to say the Monocotyledons are, in terms of the stellar analogy, representable by a number of rather large and scattered stars, which are, as it were, outliers to a relatively small central mass. This latter can only be resolved into its constituent stars at quite a different level of particulation, and thus there is in the Monocotyledons the almost paradoxical situation that the specializations, using that word in its more accurate sense, outweigh the rest. In the Dicotyledons we find quite the opposite, for here the specializations would appear not only to be less prominent qualitatively, that is in dimensions of isolation, but they are also of less overall importance quantitatively, so that it is to a much greater degree that the members of the central mass of them give characterization to the series as a whole. We may indeed say that while in the Monocotyledons it is the outlying parts of the constellation which make the greater contribution to the whole, in the Dicotyledons the weight of the constellation is much more concentrated and centralized, and this of course is only another way of repeating the earlier statement that there is much greater continuity of structure in the latter than in the former, or, to use the terms of the stellar analogy, that the stars are less widely separated from one another in the Dicotyledons than they are in the Monocotyledons.

The aptness of these comments on specialization becomes readily apparent when we examine the great mass of dicotyledons group by group, for we then become aware that these groups vary greatly among themselves in the degree to which they are particularized, and therefore unlike any others. There are some (nearly always recognized as families) which, on total characterization, are so unlike any others that they quite clearly stand apart, and may rightly be designated as isolated or detached. Some others (most of them families), have each their own overall strong features but in fact connect up with, or "run into" others of a more ordinary kind by virtue of a minority of their members, a state of affairs which may well be designated as semi-isolated or attached. Still others are separable as recognizable unities only on the basis of some particular and frequently insignificant distinction and therefore show relatively little characterization at all, and these may, in contradistinction to the rest, be thought of as non-isolated or confluent. Fewer of these are co-extensive with single taxonomic families.

Much that has been said leads to the conclusion that it will be very much easier to detect among the Dicotyledons than among the Monocotyledons what may be called an average or ordinary condition, in which there will be found the more usual run of characters expressed in their most generalized values. This is indeed so, and the elucidation of this "average dicotyledon" as it may be termed is well worth while. It must of course be realized that to obtain a theoretically perfect average every constituent character needs to be taken into account and this would scarcely be feasible, but it will suffice here to confine our attention within the ambit of the three broad features which we have already recognized as of special biological importance, namely—habit and stature; form of photosynthetic surface; and form of the inflorescence and flowers.

The first of these is the simplest, not only in the way it illustrates the nature of our quest, but also because it really involves only one variable, that of size, and there will probably be little dispute that in this respect the generalized mean condition among dicotyledons is seen in those plants which, though on the one hand having a fully woody habit, do not, on the other, possess a main axis or trunk, in short the kind of plant which is called a shrub, and whose stature, within that definition, may be thought of as somewhere between five and fifteen feet. That, at least would seem to be near the arithmetical mean and it is probably true to regard it as the most average or normal stature condition in the sense that it is of the widest taxonomic occurrence throughout the series as a whole. The importance of stature in determining the general potentiality of the vegetational environment has already been stressed, but the reader may be reminded again that the average size of flowering plants is a factor which must have played an important part in conditioning the evolution of many animals.

Foliage (or its absence) is somewhat more difficult to compare because it involves more variations, some of which are of doubtful biological significance. Thus, it may be disputed whether the rest of the biota in an environment are much affected by the difference between large simply-pinnate compound leaves with lanceolate segments and branches bearing numerous true leaves of similar shape and size, or even, indeed, between comparatively diverse leaf-blade forms. But the copiousness of foliage and the size and proximity of its blades may be of much greater importance, if only in their influence on local micro-climates. On this and other counts too it seems reasonable to recognize the mean of foliage as represented by medium-sized, more or less lanceolate, space-separated blades, which may be either simple or compound. The tegumentation of foliage is another feature with a pretty clear biological potential, and here it would seem that the generality of dicotyledons are neither distinctly hairy nor completely glabrous, but exhibit some kind of pubescence, at least in their younger parts.

Of floral form there is more to say again because of the large number of components, and as with foliage, some of these may with advantage be left aside as unlikely to be of great biological moment, as for instance the distinction between superior and inferior ovaries. Naturally it is possible to argue into any such characters a hypothetical importance, but it is essential not to overload our presentation here, and to deal only with some aspects which really contribute to the general picture.

Flowers, in general, comprise four different sorts of organs, sepals, petals, stamens and carpels, to which, with few exceptions, there is never anything truly additional, and it is familiar that different flowers may show all or less than all of these, and this in varying proportion. At one end of the scale a flower may consist of but one stamen or one carpel; at the other end it may contain all four sorts of organs, some at least in large numbers. Between these extremes there exist, as far as we can judge, a very complete series of intermediate conditions, though certain states predominate. Thus (in the Dicotyledons) if sepals are present they most commonly number five; petals, if present, do so also; stamens or carpels are not infrequently seen in much greater numbers, but it is safe to say that most often they are not more than twice as many as the sepals or petals. Equally familiar is the fact that the individual members of these different parts, or whorls, of the flower may be joined together in various ways and that there may be differences among them in shape and size, but here again it is safe to say that most commonly neither of these last possibilities is expressed. In two other particular directions also there are long variation gamuts among flowers, namely in size and aggregation. In the first of these the total range is very wide, from less than 1 mm. upwards but within it the lower values are quite clearly much more frequent than the others, so much so that it can be said that flowers measuring more than an inch or two are comparatively rare, a fact which their inevitable prominence tends, perhaps, to conceal. In the second, which is more difficult to value, it must be enough to say that variation is from large solitary flowers at one extreme to tiny flowers so closely united and co-ordinated as to have lost all individuality at the other (see Chapters 9 and 10). Finally, it is to be remembered that if either the stamens or the carpels are entirely absent from a flower or even indeed less than fully functional, there is imported into the floral scheme the additional and biologically very potent, factor of unisexuality, which, if it is spread throughout the whole body of the individual, results in the important condition of dioecism.

In each and every of these characteristics there is, within the Dicotyledons, a considerable range of variation, but it is sufficiently clear, without making elaborate mathematical calculations, that taking this huge mass of species *as a whole*, the less pronounced states of all these features predominate. We may therefore conclude for our present purposes that, as regards floral organization, the dicotyledon mean or norm is shown by those plants in which the flowers are of small to medium size (say, half an inch or rather less in dimensions); grouped in semi-compacted inflorescences; regular; hermaphrodite; dichlamydeous; pentamerous; and either mono- or diplo-stemonous. Add to this that they are borne on plants of shrubby woody habit, with open-spaced lanceolate leaf surfaces which are perhaps slightly pubescent, and we have come as near as we may reasonably hope to come to a definition of the generalized or average dicotyledonous plant.

The importance of this recognition lies in the fact that in conjunction with the three-fold categorization of groups given above it enables us already to detect not only the framework of our analysis but also the outline of the diagram by which it can best be expressed in pictorial form, for it is clear that in this latter the more average families must provide the central feature, round which the others will be grouped. In short we can, within the scope open to us here, best picture the Dicotyledons as composed of a central mass of comparatively little differentiated or confluent groups, or types as it is better to think of them (since this avoids laying down their exact limits) and peripherally of two concentric rings of groups composed respectively of those which have been referred to as *attached* and *detached*. We can also best set out the serial account of these unities, which is the next purpose of this chapter, in the same way, but in reverse, beginning with the last named category and working inwards therefrom, first to the attached groups and then to the types of the central mass. In this account the Dicotyledons are treated as far as possible as were the Monocotyledons in the last chapter, that is to say the purely taxonomic content of each of the groups is stated and each is given a generalized name, and as far as feasible this is done also for the types of the central mass.

The Isolated or Detached Groups of Dicotyledons

1. *The Composites* (*figs.* 32 and 141–162).

This group comprises the cosmopolitan family Compositae and the small South American family Calyceraceae which, biologically, differ only in relatively trivial features. This is the largest distinctive group of dicotyledons, having about 20,000 species, and raises so many important evolutionary problems that Chapter 10 is entirely devoted to some of them.

2. *The Asclepiads* (*figs*. 33 and 99–126).

These comprise the widespread but predominantly tropical family Asclepiadaceae with some 3,500 species, which is isolated structurally and biologically by its pollinial method of pollination and the associated design of the flower. This group also is dealt with more fully later, in Chapters 7 and 8.

3. *The Spurges* (*figs*. 34 and 140).

A characteristic pseudanthial group (see Chapter 9) contributing to many kinds of vegetational environment in most parts of the world, and including one of the three sorts of cactoid plants. It comprises the genus *Euphorbia* and its satellite genera of the subfamily Euphorbieae, of the family Euphorbiaceae, and has about 1,250 species.

4. *The Figs* (*fig*. 35).

A characteristic specialized group comprising the pan-tropical genus *Ficus* in Moraceae and its two satellite genera *Dammaropsis* of New Guinea and *Sparattosyce* of New Caledonia, and having some 1,250 species. The peculiar inflorescences and pollination isolate it, though its broad structural theme is not entirely unseen elsewhere in the family, as in the Olmedieae.

5. *The Podostemads* (*fig*. 36).

Perhaps the most striking example of general biological isolation, its members living on the rocks of fast-flowing streams in warm regions where there is a marked drier season and consequent temporary lowering of the water, conditions in which the thalloid vegetative form of the plants and their floral peculiarities are appropriate. They comprise the families Podostemaceae and Tristichaceae with about 200 species.

6. *The Dodders* (*fig*. 37).

Apparently the only really structurally isolated group of parasites, the peculiarity being the almost thread-like growth of the fine leafless stems. They comprise the genus and family *Cuscuta* with about 140 species in almost all parts of the world, and the rather more robust genus *Cassytha* of the Lauraceae, with about 15 species, widely spread in the tropics.

7. *The Pitcher-plants* (*fig.* 38).

These comprise the genus and family *Nepenthes* which, with about 75 very coherent species, ranges from Madagascar to Australia, and the monotypic genus and family *Cephalotus* of south-western Australia. The group is remarkably isolated in its peculiar pitcher leaves, though the two families differ somewhat in details in this respect. *Nepenthes* has characteristic dioecious flowers.

8. *The Birthworts* (*fig.* 39).

These afford the most notable instance of zygomorphy in combination with monochlamydy, and the details of their flowers, which have an unusual structure and mechanism, mark them off sharply. They comprise the 350 or so species in the large and widespread, but chiefly tropical, genus *Aristolochia* (including *Pararistolochia*) and its two small American satellites *Euglypha* and *Holostylix*, that is to say the subfamily Aristolochioideae of the Aristolochiaceae.

9. *The Marcgravias* (*fig.* 40).

An unfamiliar tropical American group comprising the family Marcgraviaceae, with about 100 species. It is isolated because some or all of the bracts of the inflorescences take the form of brightly coloured nectaries, usually associated with somewhat peculiar pollination.

10. *The Casuarinas* (*fig.* 41).

In general appearance and vegetative structure one of the most isolated groups, and having also some floral peculiarities. The group comprises the genus, family and order *Casuarina* with about 50 species native in Malaysia and Australasia.

11. *The Plantagos* (*fig.* 42).

These comprise the cosmopolitan genus *Plantago* of the family Plantaginaceae (and its satellite *Bougeria*) with about 275 species.

Familiar and therefore not very strange looking plants but having a combination of characters which in fact marks them off sharply from others.

12. *The Milkworts* (*fig.* 43).

These comprise the widespread but predominantly tropical family Polygalaceae, with about 1,000 species. They have a peculiar design of zygomorphy and there seem to be no intermediate forms connecting them with less particularized groups.

13. *The Fumarias* (*fig.* 44).

These comprise the 250 or so species of the predominantly northern temperate family Fumariaceae. Like the Milkworts they have a peculiar zygomorphy and are isolated in the same way.

14. *The Balsams* (*fig.* 45).

This group comprises the family Balsaminaceae, made up of the great genus *Impatiens*, with some 550 species, and two tiny satellites. It is essentially, though not entirely, a group of the Old World, and has a peculiar and striking design of zygomorphy, with which are often associated structural and functional features of the andrecium which enhance its isolation.

15. *The Tropaeolums* (*fig.* 46).

A group of about 80 species constituting the family Tropaeolaceae and confined to Central and South America. It has a peculiar design of zygomorphy, and is in certain respects the New World counterpart of the Balsams, though the floral structure is rather simpler.

16. *The Begonias* (*fig.* 47).

These comprise the 800 or so species, nearly all in the great genus *Begonia*, which make up the pantropical family Begoniaceae. The flowers show the rare combination of corolline monochlamydy, slight zygomorphy and monoecism, and this together with the usual asymmetry of the leaves, distinguishes them clearly. Moreover, these plants share with the Balsams and Tropaeolums a vegetative texture and consistency which is often very characteristic.

17. *The Cucurbits (fig. 48).*

These comprise the 1,000 species of the predominantly tropical family Cucurbitaceae. This family is more heterogeneous than almost any other so far mentioned, and it is difficult to define its isolation, but the general facies of most of the species, the morphology of the tendrils, features of the anatomy, and the type of fruit are all peculiar.

Summary of the isolated or detached groups

The most outstanding feature of the seventeen groups just described is the way they show that isolation may be due to a variety of causes, and that many aspects of the plant and its life may contribute partly or entirely to it. Naturally it is the less common themes of dicotyledon organization that particularly produce it, and the various groups may each be thought of as expressing some novelty or new combination in its design or function. The Composites and the Spurges are widely different and original expressions of the pseudanthial theme; four, namely the Podostemads, the Pitcher-plants, the Dodders and the Casuarinas, show originality which is more particularly vegetative, associated in two of them with abnormal methods of nutrition; six show particular forms of zygomorphy, complicated in the Begonias by monoecism, while of the other five two are spurred types and three are not; the Asclepiads have, among dicotyledons, a unique kind of floral mechanism and structure, and so to some degree have the Figs, while the Marcgravias show a unique state of affairs in the inflorescence. Finally two, the Plantagos and the Cucurbits, show that isolation may be due not so much to the pressing of a simple theme as to departure from the usual on a wider if shallower front. Thus, although the isolation of the former is not easily described it is, by and large, as marked as in almost any other group and very much the same is true of the Cucurbits.

The groups vary greatly in size, the total being about 31,000 species. Of these the Composites account for more than 60 per cent. and the Asclepiads for more than 10 per cent. On the other hand nine small groups together contribute only about 5 per cent.

The Semi-Isolated or Attached Groups of Dicotyledons

1. *The Proteads (fig. 49).*

This group comprises those striking and mainly African members of the family Proteaceae in which the flowers are organized into pseudanthia and which number about 500 species. These connect however with the rest of the family, which are not so remarkable, by various intermediates.

2. *The Dorstenias (fig. 50).*

This group is made up of the almost entirely American and African tropical genus *Dorstenia* of the family Moraceae, with about 150 species. The green expanded inflorescences of these usually herbaceous plants give them a very characteristic appearance but scarcely isolate them completely from other genera of the same and similar families, although they have a very unusual method of fruit discharge.

3. *The Catkin-bearers (fig. 51).*

These woody plants are a prominent feature of flowering plant vegetation especially outside the tropics, but they cannot be regarded as isolated because they contain both monoecious and dioecious types and because there are many intermediates between catkins and other inflorescences, especially in the Ulmaceae and Moraceae. They form a large and ecologically important almost world-wide group of some 1,750 species, and comprise the families Balanopsidaceae, Betulaceae, Corylaceae, Fagaceae, Garryaceae, Juglandaceae, Leitneriaceae, Myricaceae, Myrothamnaceae and Salicaceae.

4. *The Nutmegs (fig. 52).*

Fully dioecious plants are represented in the temperate regions so very largely by the Catkin-bearers that it is difficult to find a familiar name to cover the many other dioecious woody plants which are common in warm countries and there has been pressed into service the name of the best-known member of the family in which dioecism of this more general sort is most invariable, the Myristicaceae. Included in this group also are the remarkably few herbaceous dioecious plants and the rather intermediate, climbing, Menispermaceae. All these connect up with the central mass of dicotyledons by members of such families as the Sapindaceae and Ebenaceae in which individuals may or may not be fully dioecious. On the other hand the sexual state of dioecism is biologically so different from any other that these Nutmegs must be regarded as a semi-isolated group, which is widespread though predominantly tropical and which has about 3,500 species. It comprises members of many families, among the most important contributors being Myristicaceae (all), Menispermaceae (all), Octoknemataceae (all), Atherospermataceae, Elaeagnaceae, Euphorbiaceae (which provide more than half the total), Flacourtiaceae, Monimiaceae, Moraceae, Santalaceae, Thymelaeaceae, Phytolaccaceae and Ulmaceae.

5. *The Mangroves* (*fig. 53*).

These may be regarded as the ecological counterpart, among the woody dicotyledons, of the marine angiosperms in the Monocotyledons, but since they are subaerial plants they afford a very special kind of vegetational environment for other biota. The most characteristic of them show aerating roots and viviparous germination, and it is to these, about 30 tropical species, that the group is more particularly restricted. They belong to some ten genera in Rhizophoraceae, Combretaceae, Sonneratiaceae, Myrsinaceae, Verbenaceae and Pellicieraceae.

6. *The Mistletoes* (*fig. 54*).

These comprise the large (1,300 species) and almost worldwide, though chiefly tropical family Loranthaceae, the members of which are semi-parasitic on woody hosts, but there must also be included here, if only for convenience, one or two genera of Santalaceae, among them *Phacellaria*, which actually is parasitic on species of Loranthaceae, and the small South American genus and family, *Myzodendron*. The few temperate members of the group are less striking than the tropical and the whole connects clearly, both structurally and functionally, with the non-pseudanthial Proteaceae through such genera as *Nuytsia*, which is said to be quite autotrophic.

7. *Other Parasitic Dicotyledons* (*fig. 55*).

It is of great evolutionary interest that parasitic flowering plants, though all dicotyledons, are of many sorts and relationships, and it

is only under pressure of space that those not already mentioned
are grouped together here. In very general terms they comprise
three kinds of plants.

1. Those which differ but little from more normal plants except
in the absence of chlorophyll and of which the most important are
the families Orobanchaceae, Monotropaceae and Lennoaceae;
certain species of Gentianaceae; and the genera *Cytinus* and
Bdallophyton.

2. Those with very small and numerous flowers and usually
bizarre vegetative form, comprising the families Balanophoraceae
and Cynomoriaceae.

3. Those which are wholly haustorial in the vegetative state, and
often with very large flowers, the Rafflesiaceae and Hydnoraceae.

If the relatively very few similar but saprophytic dicotyledonous
plants are included here also the total is about 350 species and the
distribution world-wide. Sundry semi-parasitic plants connect this
group with the central mass.

8. The Crucifers (*fig. 56*).

These comprise the great and cosmopolitan family Cruciferae
(2,500 species) which, though so generally characteristic in
appearance, nevertheless connects up by a few of its genera with
the more heterogeneous family Capparidaceae in the central mass
(see the reference to double stars on p. 60).

9. The Umbellifers (*fig. 57*).

This group has many parallels with the Crucifers, and comprises the essentially temperate and herbaceous family Umbelliferae, with 3,000 species, which connects up with the more tropical
and predominantly woody family Araliaceae in the central mass
(also see p. 60).

10. The Droserads (*fig. 58*).

The dichlamydeous insectivorous dicotyledons have been
treated here as forming two groups, one with regular and the other
with zygomorphic flowers, and the Droserads are the first of these,
comprising the more or less cosmopolitan family Droseraceae,
with about 90 species, and the American family Sarraceniaceae
with 10 species.

11. *The Eucalypts* (*fig.* 59).

This group is composed of the genus *Eucalyptus*, with 300 species, of the family Myrtaceae, and is found in Australia and eastern Malaysia. It provides perhaps the most important single element in the Australian flora. Though in detail quite strongly delineated, especially perhaps vegetatively, the genus broadly connects up with others of its family.

12. *The Acacias* (*fig.* 60).

This group includes those members of the family Mimosaceae, broadly the tropical genera *Acacia* (chiefly in the Old World) and *Mimosa* (chiefly in the New World), in which the flowers are closely aggregated in various ways and effectively little more than tiny bunches of stamens. It contains about 1,250 species, and runs into other parts of the family, notably the *Calliandra* type, which are in the central mass.

13. *The Peas* (*fig.* 61).

This huge cosmopolitan assembly, the family Papilionaceae, of more than 10,000 species, is a good example of a semi-isolated group because, although the great majority of its members are characteristically enough zygomorphic, it connects by a minority of them, notably those of the Sophoreae, with such regular-flowered members of the central mass as some of the Rosaceae and Connaraceae.

14. *The Cassias* (*fig.* 62).

Although this group, which comprises most of the family Caesalpiniaceae, with about 1,800 species, differs from the Peas in the details of its zygomorphy and in being more exclusively tropical, the two are parallel in many ways. It connects with the central mass through such of its regular-flowered members as the genera *Haematoxylon* and *Vouacapoua*.

15. *The Lobelias* (*fig.* 63).

These comprise the widespread family Lobeliaceae, together with the two predominantly Australian families Goodeniaceae and Stylidiaceae, and contains about 1,250 species in all. There are notable distinctions between these three but there are also significant overall similarities, especially in their zygomorphy and other floral details, which bind them together strongly. At the same time they link up with the central mass through some of their simpler-flowered and almost or quite regular-flowered members, and by the free-stamened species of the genus *Cyphia*.

16. *The Gesneriads* (*fig.* 64).

The characteristic zygomorphy of this group, which is well termed declinate or bent down is perhaps the simplest sort, and when but slightly expressed connects up with many regular flower types. It is well illustrated by the familiar foxglove (*Digitalis*) and various other Scrophulariaceae, but is more particularly to be seen, though by no means invariably, in the more tropical families Gesneriaceae, Bignoniaceae, Verbenaceae, Pedaliaceae and Martyniaceae. It is therefore rather difficult to estimate its numerical value but 2,000 species is probably about the right figure.

17. *The Labiates* (*fig.* 65).

The zygomorphy here is of the bilabiate or personate design seen most markedly, though not exclusively, in the world-wide family Labiatae and the tropical family Acanthaceae, and less generally among Gesneriaceae, Scrophulariaceae and Verbenaceae. In many ways it pairs with the declinate type and in fact there are strong links between them, though the typical members of each are very distinct. The Labiates, which probably number some 10,000 species, connect with the central mass through those which are almost or quite regular-flowered.

18. *The Lentibulariads* (*fig.* 66).

These comprise the zygomorphic dichlamydeous insectivorous plants, namely the almost world-wide family Lentibulariaceae (300 species). Their floral design is labiate but their notable vegetative peculiarities justify their separation. They may be regarded as linking up with the central mass through some of their smaller and least heterotrophic members.

19. *The Cupheas* (*fig.* 67).

These comprise the almost entirely tropical American genus *Cuphea*, with about 230 species, and its tiny satellite *Pleurophora*, in the family Lythraceae, which show a particular short-spurred kind of zygomorphy associated with variation in petal development, and this strongly characterises them. They link with the central mass through the regular-flowered members of their family.

20. *The Violets* (*fig.* 68).

These again provide a typical and familiar semi-isolated group, comprising the zygomorphic members of the widespread family Violaceae, amounting to about 500 species, mostly in the herbaceous genus *Viola*. They connect with the central mass through regular-flowered members of the family.

21. *The Passion-flowers* (*fig.* 69).

The familiar name of this group recalls its floral peculiarities, which are often accompanied by unusual leaf-shapes, but the former are only fully seen in the tropical genus *Passiflora* in the wide sense, with 400 species, to which the group should therefore be limited. Other genera of the Passifloraceae merge into such families as Achariaceae and Malesherbiaceae in the central mass.

22. *The Mesembryanths* (*fig.* 70).

In this group which comprises the most typical members of the great genus *Mesembryanthemum* in its widest sense, in the family Aizoaceae, there is often extreme leaf succulence and great multiplication of petal-like staminodes in flowers which are essentially large and more or less solitary but some of the species are less notable and connect, through other genera, with the central mass. *Mesembryanthemum* in this wide sense has some 800 species, the very great majority of them in South Africa.

23. *The Cacti* (*fig.* 71).

This group comprises the third and last large assembly of cactoid or stem-succulent flowering plants, which, unlike the other two (see pp. 125, 225, 339) has its own kind of flower, characterized by a multiseriate perianth, polystemony, often some slight zygomorphy, and a rather particular texture. It forms the American (predominantly tropical) family Cactaceae, with perhaps 1,250 species. The family includes at least one genus however which is truly leafy, and through this and some of the simpler-flowered members links up with the central mass.

24. *The Water-lilies (fig. 72).*

The combination of floating leaf blades and large, usually floating, flowers in this group, which comprises the world-wide family Nymphaeaceae in its stricter sense, with about 60 species, makes it a striking one but does not isolate it entirely from members of other families with similar vegetative form or floral organization.

Summary of the semi-isolated or attached groups

As with the isolated groups, the semi-isolated are by no means all of one kind, and up to a point there is an interesting parallel between the two series. Thus, in the latter the Crucifers and the Umbellifers are the counterparts of the Composites ecologically and of the more ordinary Plantagos structurally, while the set of zygomorphic groups parallel the Milkworts, Balsams and others in the former. The Passion-flowers correspond in a general way with the Cucurbits and Begonias, and the Mangroves and the Water-lilies may be thought of as corresponding to the Podostemads. There are moreover pseudanthial, dioecious, parasitic and insectivorous groups in each. On the other hand there are not, among the semi-isolated, any extraordinary pollination types such as the Asclepiads and Figs or, in lesser degree, the Marcgravias. Conversely there is nothing in the isolated series which fully parallels the succulent Mesembryanths or Cacti.

The total species numbers in the semi-isolated groups is about 42,000, or about one-third again as many as in the isolated.

The Central Mass of Dicotyledons

Although the isolated and semi-isolated groups have, between them, accounted for many thousands of dicotyledon species, including some of the largest taxa, what still remains amounts, in species reckoning, to about two-thirds of the whole, and this larger portion is, by the nature of the case, more closely knit and therefore more difficult to analyse. So much so indeed that we shall be justified in using any reasonable representational convention which will help towards simplifying the picture. Such a convention is to visualize the central mass as composed of an inner part, or core, and an outer part, or periphery. The former contains the most generalized types, conforming to our description of the average dicotyledon as a shrubby woody plant with medium-sized hermaphrodite dichlamydeous oligostemonous flowers in well-defined but not congested inflorescences (namely such a plant as is familiar in laurustinus, *Viburnum tinus*, or in lilac, *Syringa vulgaris*), and the latter contains the remainder, which may be regarded as lying between the core and the semi-isolated and isolated groups. Just exactly where the limits of the core are drawn must inevitably be a matter of opinion, but agreement is not a matter of consequence and our definition provides sufficient precision. In proceeding to the details behind and within this simple representation it is convenient to consider the simpler aspect, that of the core, first.

The Core of the Dicotyledons

Bearing in mind the definition just given the range of structure within the core is surprisingly wide, especially in respect of inflorescence and leaf form, in which, for example, the differences between racemose and umbellate and between entire and divided may be the bases of considerable visual differentiation. It would be manifestly impossible to refer to all its inflections, and all that can be done here is to illustrate them by listing the families which are the larger contributors to the core, leaving the reader to refer to other sources for more extended accounts of them. These families tend to fall into two sets, though it would be idle to attempt a rigid separation of them, and the first of these, which must surely be regarded as comprising the most generalized of all dicotyledons, includes the Aceraceae, Anacardiaceae, Aquifoliaceae, Burseraceae, Celastraceae, Connaraceae, Erythroxylaceae, Hippocrateaceae, Myrsinaceae, Olacaceae, Rhamnaceae, Sabiaceae, Sapotaceae and Simaroubaceae, together with some other smaller families or part of families. In the second set the flowers are on the whole larger and more conspicuous, and here the chief families are Ehretiaceae, Elaeocarpaceae, Epacridaceae, Escalloniaceae, Grossulariaceae, Hamamelidaceae, Hydrangeaceae, Icacinaceae, Meliaceae, Pittosporaceae and Vacciniaceae, to which must be added very considerable proportionate contingents from the Apocynaceae, Caprifoliaceae, Ericaceae, Loganiaceae, Oleaceae, Rubiaceae, Rutaceae and Sterculiaceae.

Though these plants conform so closely to the average many of them, especially in the latter set, have some feature or combination of features which, but minor in absolute dimensions, may nevertheless convey a very distinctive appearance. To mention only two of many possible examples, the Sapotaceae have particularities of leaf design and arrangement, and of grouping and shape of flowers, which gives most of them an unmistakable facies, while in many of the broader leaved Ericaceae and Vacciniaceae the particular leaf texture and the posture and form of the waxy urceolate flowers (fig. 96) do the same. Again, dissection of the leaves, such as is seen in the palmatifid Aceraceae (fig. 93), often gives considerable character.

It is difficult to calculate the size of the core because so many families make at least some contribution to it, but it would seem, on the basis of the families mentioned, to contain something of the order of 20,000 species. It is perhaps worth noting that no one of the largest families falls entirely within it. Representative types here are *Acer, Sambucus, Syringa* and *Pieris* (figs. 93–96).

The Peripheral Part of the Central Mass

When, as can now be done, the members of the core are compared with the members of the outlying groups earlier described, it is readily realized that the less markedly un-average plants which constitute the peripheral part of the central mass, are mainly composed of types showing incipient states of the specializations so prominent in those groups. If we may be allowed the phrase without any directional evolutionary implications, and merely as a means of pointing comparisons, it may be said that there are various biological and structural trends away from the average, some of these greatly preponderating, and these may be thought of as providing distinctive segments of the periphery of the central mass. Sometimes there is a combination of trends which cuts across these segments, but these are on the whole few. In this way it is possible to recognize five segments, though these are very different in size.

First is a large segment of herbaceous plants containing all those types which differ from the average mainly or only in having herbaceous instead of woody life-forms. They comprise all, or all but the larger-flowered members of the Boraginaceae, Caryophyllaceae, Crassulaceae, Gentianaceae, Hydrophyllaceae, Linaceae, Oxalidaceae, Pirolaceae, Plumbaginaceae, Polemoniaceae,

Primulaceae, Saxifragaceae and certain smaller families, as well as contingents from many others, especially the Araliaceae, Rubiaceae and Solanaceae. Representative types of this segment are *Cotyledon*, *Gypsophila* and *Symphytum* (figs. 73, 74, 75), and its position is clearly between the core and such groups as the Crucifers and Umbellifers.

The second segment contains those types in which the flowers are smaller and of simpler structure than the average. This is a large and varied segment, partly because it comprises both woody and herbaceous plants and partly because marked simplicity of floral organization often involves unisexuality, and it obviously lies between the core and the small series of groups of which the Catkin-bearers and the Nutmegs are the largest. The chief contributors are all or part (i.e. the hermaphrodite or monoecious non-pseudanthial members) of the Amaranthaceae, Chenopodiaceae, Euphorbiaceae, Illecebraceae, Lauraceae, Monimiaceae, Moraceae, Nyctaginaceae, Oleaceae, Piperaceae, Polygonaceae, Proteaceae, Santalaceae, Thymelaeaceae, Urticaceae, Vitaceae (effectively) and a few small families. Representative peripheral types here are *Artocarpus*, *Chenopodium*, *Banksia*, *Piper* or *Peperomia* and *Vitis* or *Parthenocissus* (figs. 76–80).

The third segment comprehends two rather different and opposing trends which commonly merge into one another, namely increased corolla size and increased number *or* prominence of stamens. It would seem that there are here involved some of the more mechanical factors in floral design, such as, for instance, express themselves in a tendency for stamens to be more numerous rather than individually larger and to be less variable in size than other parts of the flower, considerations which no doubt bear on problems of cell size and pollination. Suffice it to say that the types of this segment concentrate in three main designs—the large oligostemonous flower e.g. *Convolvulus*; the large polystemonous flower e.g. *Paeonia*; and the smaller polystemonous flower in which the stamens, because of their number or prominence, conceal the rest of the flower e.g. *Callistemon*. A woody habit predominates in this segment but all these conditions occur among its numerous herbaceous members.

There is only one considerable family, the Convolvulaceae, in which there is a high proportion of large oligostemonous-flowered plants, the other chief contributors to it being the Apocynaceae, Onagraceae, Rubiaceae, and Solanaceae. It is of considerable interest that the types most notable for sheer size of flower, such as the Birthworts, Cucurbits, Rafflesiaceae and Stapelieae (see next chapter), are all parts of isolated or semi-isolated groups, so that extreme flower size is generally associated with other peculiarities. The larger-flowered polystemonous types on the other hand provide the great bulk of the segment as will be apparent from the list of families which totally or largely contribute to them, namely the Actinidiaceae, Annonaceae, Bombacaceae, Capparidaceae, Chlaenaceae, Cistaceae, Dilleniaceae, Dipterocarpaceae, Flacourtiaceae, Guttiferae, Hypericaceae, Lecythidaceae, Loasaceae, Magnoliaceae, Malvaceae, Myrtaceae, Ochnaceae, Papaveraceae, Portulacaceae, Ranunculaceae, Rosaceae, Rutaceae, Sauraujaceae, Symplocaceae, Theaceae, Tiliaceae. The polystemonous small-flowered types again provide but a small part of the segment, especially because their more extreme expressions have been recognized already as the semi-isolated group of the Acacias. One or two families are totally or mostly included here, notably the Combretaceae and Cunoniaceae; the Myrtaceae contribute strongly to it; and various of the families above afford examples, but otherwise it is largely an affair of the remaining members of the Mimosaceae. Representative types in this third segment are *Convolvulus* or *Calystegia*, *Fuchsia*, *Hibiscus*, *Magnolia*, *Pyrus*, *Eugenia* and *Calliandra* (figs. 81–87).

The fourth segment contains those plants of the central mass which differ from the average in showing some kind and degree of zygomorphy. This is a prominent general feature of the Flowering Plants and many designs of it are so copiously expressed that they have already been accounted for among the groups, but there are many other less pervasive examples which can hardly be extracted from the central mass without unduly multiplying the groups. These are best considered under three heads.

To begin with there are sundry taxonomically scattered and rather peculiar types, which, though approaching some of the groups can hardly be included therein, and of these *Schizanthus* in Solanaceae, and *Delphinium* and *Aconitum* in Ranunculaceae are good examples. Then there are a few patterns sufficiently expressed to be segregated taxonomically which are well exemplified by the families Resedaceae and Valerianaceae. Lastly, there are the more numerous types in which the floral irregularity is totally or very largely a matter either of the andrecium or the andrecium and gynecium only. This is apparently little seen outside the tropics, but there three families serve to illustrate it well. In the Chrysobalanaceae irregularity in the stamens is various and is accompanied by asymmetry of the gynecium, for the reason that this consists of a single carpel with a basal style. In some members of the Lecythidaceae one segment of the staminal ring is much enlarged and folded inwards to form a hood over the gynecium. In the large family Melastomataceae the andrecium is again irregular and of a rather special kind, and this is accompanied very generally by peculiarities of foliage, the combination giving these plants a most characteristic facies. Representative types here are *Schizanthus*, *Reseda* and *Melastoma* (figs. 88–90).

The fifth segment contains those peripheral types of the central mass which, on account of some special peculiarity or combination of more ordinary characters make so much their own biological contribution to any environment in which they occur that they cannot be left undistinguished. What these exactly are must naturally be much a matter of personal opinion and they are also likely to be very heterogeneous, and it is best therefore merely to illustrate them by one or two token examples. One of the most important is undoubtedly that of the microphyllous "ericoid" plants, illustrated by *Calluna* (fig. 91), which because of their particular woody life form, their xeromorphic foliage, and, frequently, their gregariousness, tend to afford unique environmental conditions. These plants belong to sundry taxa besides the Ericoideae, and their floral characters are various. A very different type is that of such completely or almost completely submerged aquatics as *Ceratophyllum* or *Myriophyllum* (fig. 92), *Elatine* (fig. 92) and *Callitriche*.

These peripheral segments of the central mass contain at least half the Dicotyledons and probably about 100,000 species.

Conspectus

The aim of these last three chapters has been to give some impression of the Flowering Plants without undue emphasis on their more purely taxonomic classification, and in order to illustrate the value of the opinion that a proper realization of what these plants are is an essential preliminary to any consideration of or deduction as to how they may have come into being. The first of the three chapters dealt with some general aspects of these plants and their presentation, and introduced some of its particular problems. The second chapter dealt with the Monocotyledons as the smaller and more easily analysed of the two main series of flowering plants, and there in conclusion it was found possible to construct, and reproduce in graphic form, a three-dimensional model depicting the constituent groups to scale as spheres with diameters proportionate to the cube roots of their species numbers, and in their relative positions according to their structural characteristics.

In this, third, chapter the more complex series of the Dicotyledons has been dealt with, in rather a different way because of its greater taxonomic size and closer texture. There have been recognized, chiefly in terms of floral structure, first a sequence of what have been called isolated, or detached, groups; second a sequence of what have been called semi-isolated, or attached, groups: and third a central mass of relatively more generalized types. This has been illustrated by a simple two-dimensional scale diagram (fig. 97) in which the various unities are represented by circles of which the diameters are proportional to the square roots of their species numbers. The central mass was then dissected and its constitution found to be expressible in terms of a core, comprising the most average types of Dicotyledons, and a number of peripheral segments, each in the main illustrating one major organizational theme connecting the core with certain of the semi-isolated groups. This dissection is illustrated in a second diagram (fig. 98).

FIG. 32. A Composite: *Calendula officinalis,* × 1½, after Baillon

FIG. 33. An Asclepiad: *Hoya carnosa*, slightly reduced, partly after Nicholson

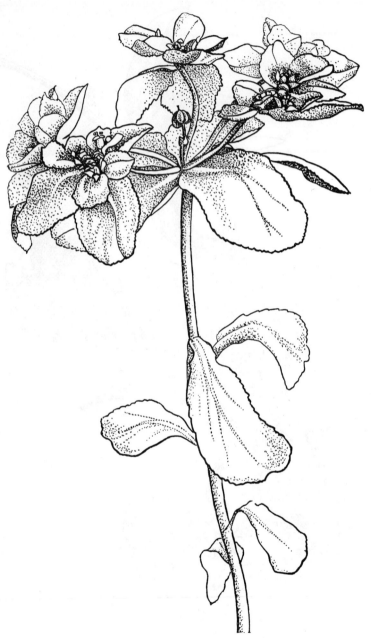

FIG. 34. A Spurge: *Euphorbia helioscopia*, slightly enlarged

FIG. 35. A Fig: *Ficus carica,* × $\frac{2}{5}$

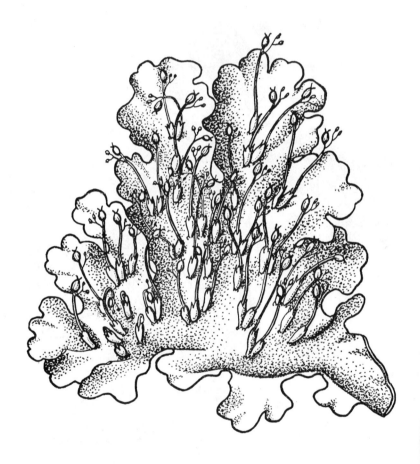

FIG. 36. A Podostemad: *Inversodicraea pellucida*, slightly enlarged, after *Pflanzenfam*.

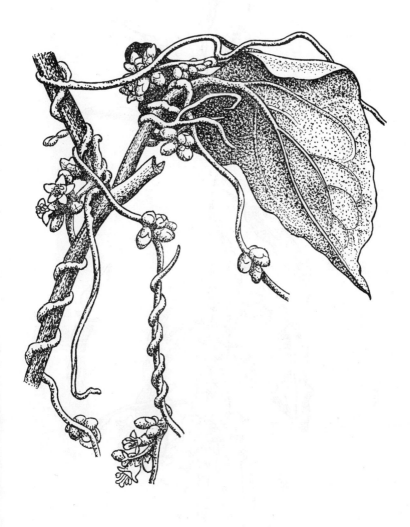

FIG. 37. A Dodder: *Cuscuta gronovii*, × 1½, after Bailey

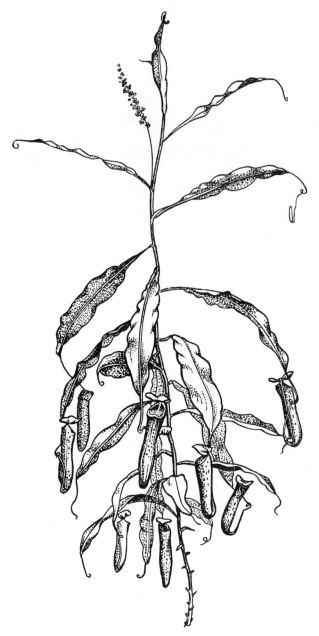

FIG. 38. A Pitcher-plant: *Nepenthes distillatoria*, × ⅛, after Nicholson

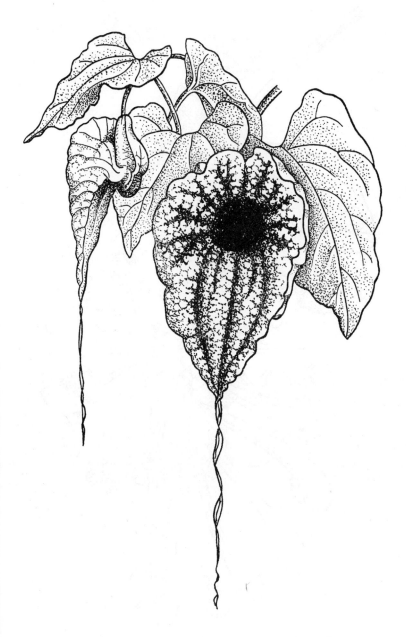

Fig. 39. A Birthwort: *Aristolochia grandiflora*, × ⅙, after Bailey

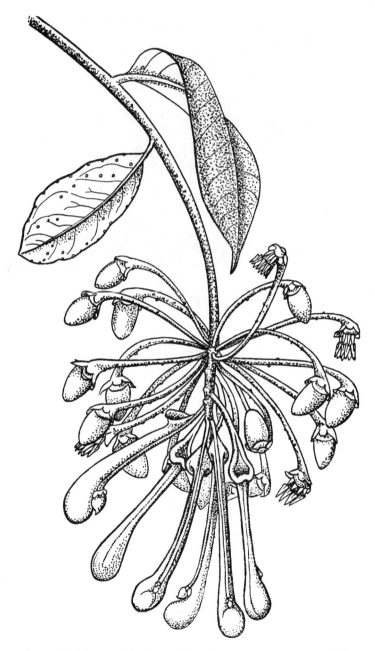

FIG. 40. A Marcgravia: *Marcgravia roraimae,* natural size, after Engler

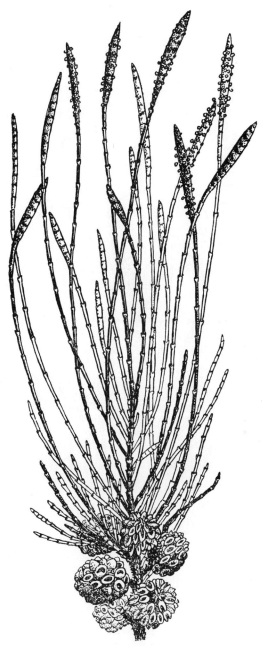

FIG. 41. A Casuarina: *Casuarina equisetifolia*, natural size, after *Pflanzenfam*.

FIG. 42. A Plantago: *Plantago major*, natural size

FIG. 43. A Milkwort: *Polygala chamaebuxus*, slightly enlarged, after Bennett

FIG. 44. A Fumaria: *Dicentra spectabilis*, × ⅓, after Bailey

FIG. 45. A Balsam: *Impatiens glandulifera*, slightly reduced

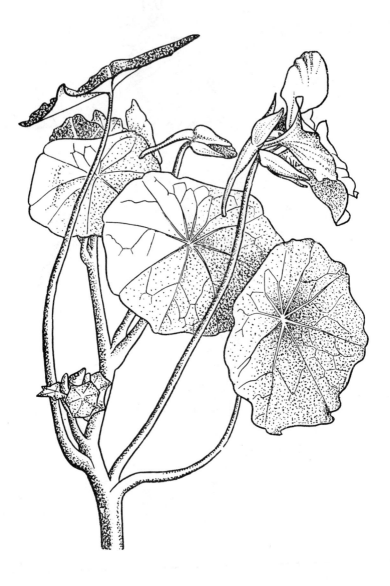

FIG. 46. A Tropaeolum : *Tropaeolum majus*, natural size

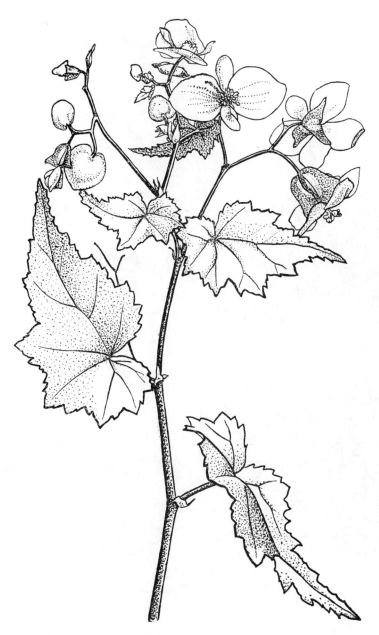

FIG. 47. A Begonia: *Begonia* sp., natural size

FIG. 48. A Cucurbit: *Cucurbita pepo*, × ¼, after Baillon

Fig. 49. A Protead: *Protea cynaroides*, × ⅓, various sources

FIG. 50. A Dorstenia: *Dorstenia mannii*, × ⅔, after *Bot. Mag.*

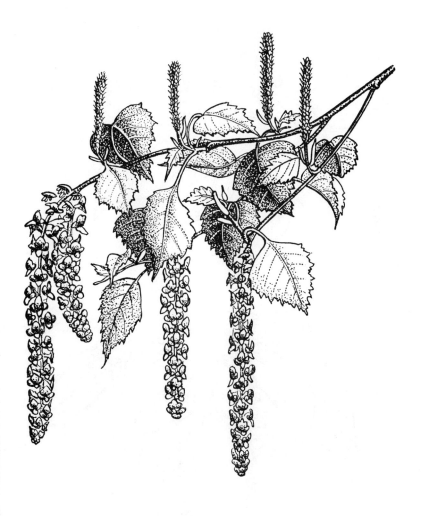

FIG. 51. A Catkin-bearer: *Betula alba*, natural size

FIG. 52. A Nutmeg: *Myristica fragrans*, slightly reduced, after Lotsy and Baillon

FIG. 53. A Mangrove: *Rhizophora* sp., somewhat reduced, after Baillon

FIG. 54. A Mistletoe: *Psittacanthus dichrous*, × ½, after *Pflanzenfam*.

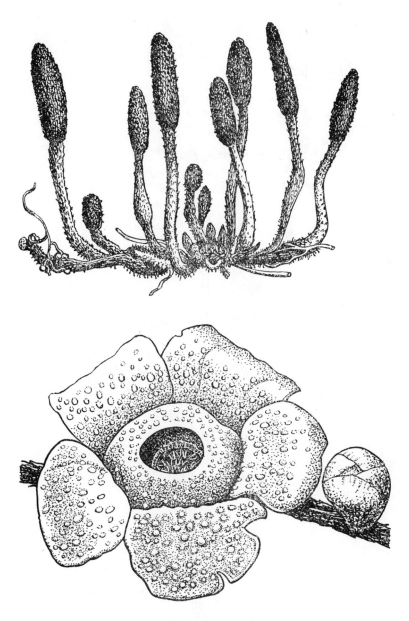

FIG. 55. Other Parasitic Dicotyledons: *Cynomorium coccineum*, × ¼, after Le Maout and Decaisne, and (below) *Rafflesia* sp., × ⅛, various sources

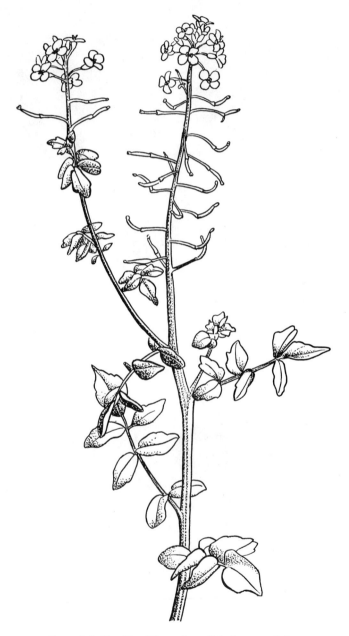

FIG. 56. A Crucifer: *Nasturtium officinale*, natural size

FIG. 57. An Umbellifer: *Conium maculatum*, × ½, after Baillon

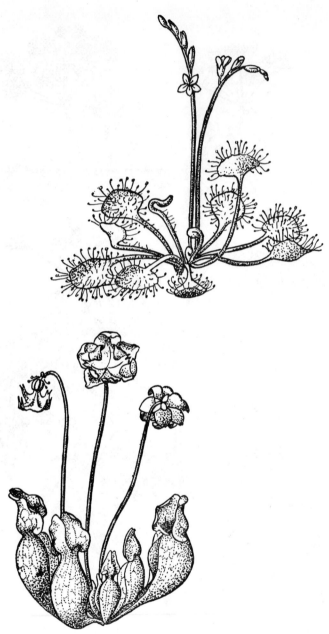

FIG. 58. Droserads: *Drosera rotundifolia*, natural size, and (below) *Sarracenia* sp., × ⅓, after Brown

FIG. 59. A Eucalypt: *Eucalyptus globulus*, $\times \frac{1}{2}$, partly after *Pflanzenfam.*

FIG. 60. An Acacia: *Acacia arabica*, natural size, after Baillon

FIG. 61. A Pea : *Laburnum anagyroides,* slightly reduced

Fig. 62. A Cassia: *Cassia obovata*, × ½, after Baillon

Fig. 63. A Lobelia: *Lobelia cardinalis*, natural size

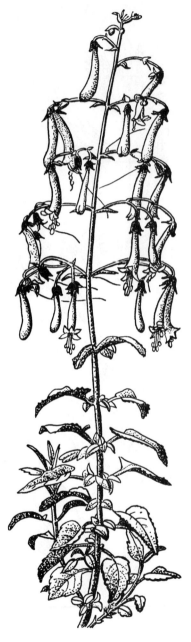

FIG. 64. A Gesneriad: *Phygelius capensis*, × ½, after Nicholson

FIG. 65. A Labiate: *Stachys sylvatica*, × ½

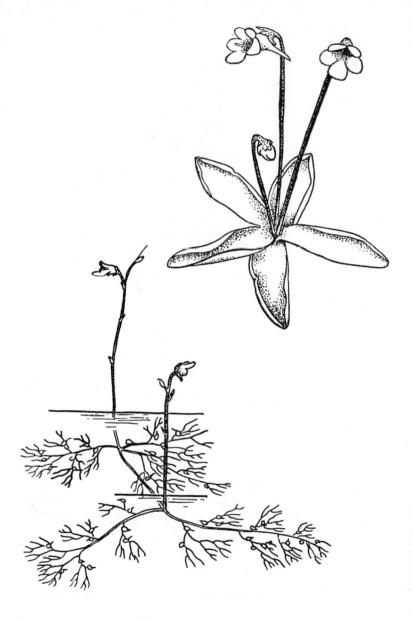

FIG. 66. Lentibulariads: *Pinguicula vulgaris,* natural size, various sources, and (below) *Utricularia minor,* slightly reduced, chiefly after Kerner and Oliver

FIG. 67. A Cuphea: *Cuphea ignea*, natural size

FIG. 68. A Violet: *Viola cornuta*, natural size, after *Bot. Mag.*

FIG. 69. A Passion-flower: *Passiflora caerulea*, slightly reduced, after Baillon

Fig. 70. A Mesembryanth: *Mesembryanthemum (Lampranthus) spectabile,*
natural size, after *Bot. Mag*

FIG. 71. A Cactus: *Echinocactus joadii*, natural size, after *Bot. Mag.*

Fig. 72. A Water-lily: *Nymphaea alba*, × ½

FIG. 73. *Cotyledon* sp., $\times$ $\frac{2}{3}$, chiefly after Bailey

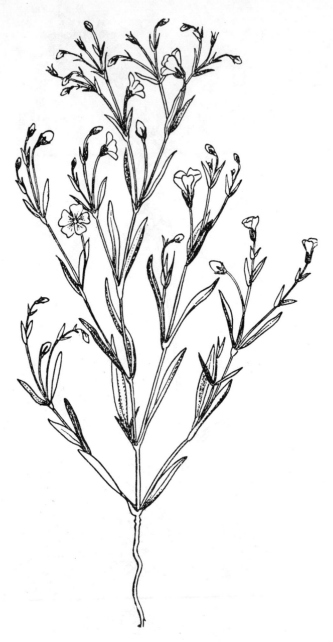

FIG. 74. *Gypsophila muralis*, slightly enlarged, after Hegi

FIG. 75. *Symphytum officinale*, natural size

FIG. 76. *Artocarpus incisa*, $\times \frac{1}{4}$, after Baillon

FIG. 77. *Chenopodium album*, natural size

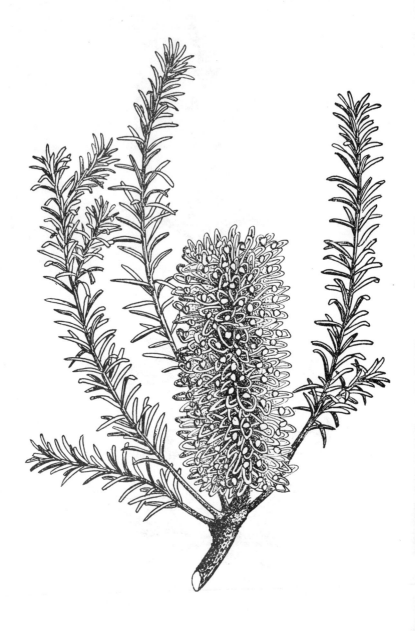

Fig. 78. *Banksia ericifolia*, slightly reduced, after Baillon

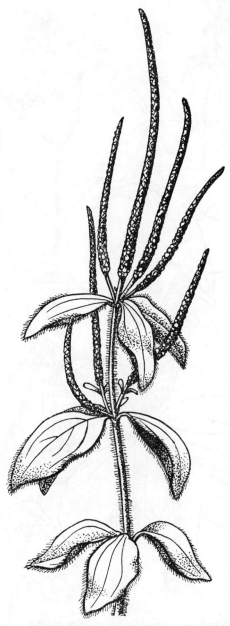

FIG. 79. *Peperomia blanda*, natural size, after Baillon

FIG. 80. *Parthenocissus quinquefolia*, natural size, chiefly after *Pflanzenfam.*

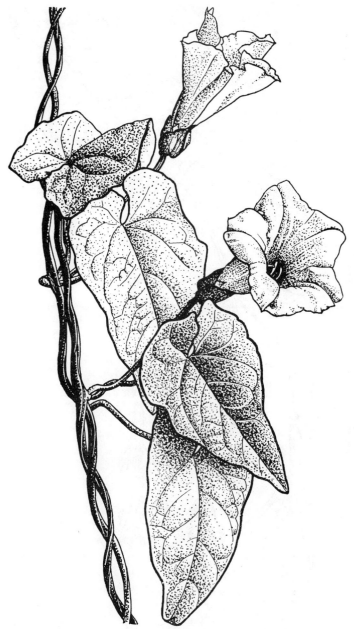

FIG. 81. *Calystegia sepium*, slightly reduced

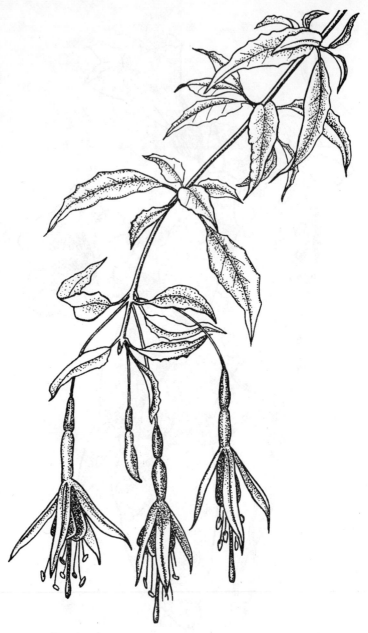

FIG. 82. *Fuchsia decussata*, natural size, after *Bot. Mag.*

FIG. 83. *Hibiscus rosa-sinensis*, natural size, after *Bot. Mag.*

FIG. 84. *Magnolia grandiflora*, × ½, after Baillon

FIG. 85. *Pyrus communis*, natural size, after Baillon

FIG. 86. *Eugenia caryophyllata*, natural size, after Baillon

FIG. 87. *Calliandra houstoni*, natural size, after Edwards in *Bot. Reg.*

FIG. 88. *Schizanthus pinnatus*, natural size

FIG. 89. *Reseda luteola*, × ½, after Baillon

FIG. 90. *Melastoma malabathricum*, natural size, after Baillon

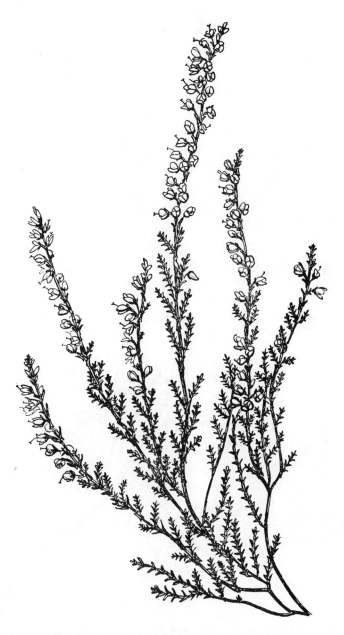

FIG. 91. *Calluna vulgaris*, natural size

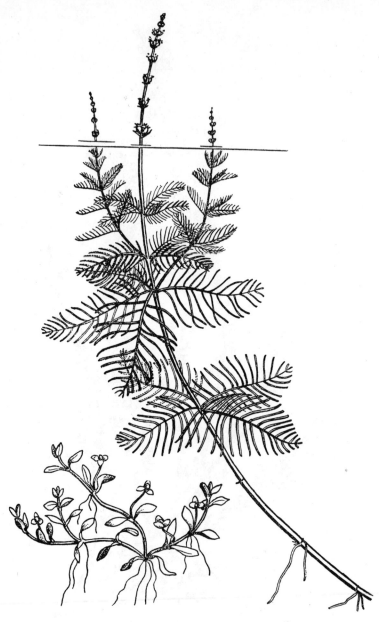

FIG. 92. *Myriophyllum spicatum*, after Fitch and Smith: *Elatine hexandra*
(below), after Sowerby, both natural size

FIG. 93. *Acer pseudoplatanus*, slightly reduced

FIG. 94. *Sambucus nigra*, natural size

FIG. 95. *Syringa vulgaris*, natural size

FIG. 96. *Pieris* sp., slightly reduced, chiefly after Bailey

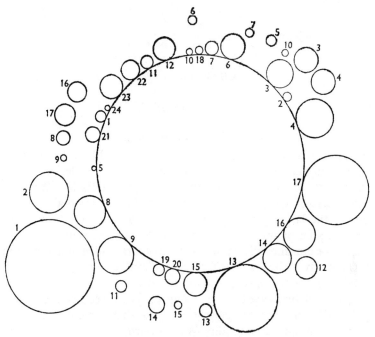

FIG. 97. A Diagrammatic Representation of the Dicotyledons as a whole
(see p. 141)

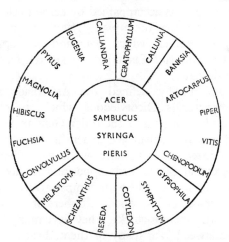

FIG. 98. A Diagrammatic Representation of the Central Mass of the
Dicotyledons (see p. 141)

CHAPTER SEVEN

THE ASCLEPIADACEAE—I

General Consideration

THE larger component groups of the Flowering Plants having now been reviewed, we must go on to enquire, by considering one of them in more detail, how evolution has expressed itself *within* such larger constituent parts. As a subject for investigation almost any one of the larger groups would be suitable and would doubtless reveal much of interest, but for various reasons some are more useful for our immediate purpose than others, and among these the great family of the Asclepiadaceae stand out pre-eminently because it is particularly well-defined and detached.

The Asclepiads are unique among the Dicotyledons in having a pollination mechanism closely akin to that which makes the Orchids unique among the Monocotyledons, and the parallel between these two in this respect is one of the most remarkable in the Flowering Plants. This alone endows the Asclepiads with special interest but it is accompanied by certain other characters, and in such fashion, that these plants reveal in a peculiarly vivid way the kind of problem which, over and over again, presents itself when the Flowering Plants are considered from the evolutionary point of view.

Formally and briefly defined the Asclepiads are regular-flowered gamopetalous Dicotyledons which, in classifications, usually find a place in or near the order Gentianales. They have been divided into more than 300 genera and 3,500 species, and only about a dozen other families are larger. They are almost invariably perennial plants with opposite and decussate entire exstipulate simple leaves, and with comparatively few exceptions they are either more or less woody climbers or erect suffruticose herbs, often with tuberous stocks. They have milky latex, whence the name "milkweed" for many of them. Features which contribute especially to the characterization of the flowers are the generally contorted aestivation of the corolla, the presence of a corona (which is often complex) and the consolidation of the andrecium and gynecium into a gynostegium. They have a particular form of follicular fruit, derived from an ovary of two carpels which, in the flower, are free below but united above, and there are many silky-plumed seeds. They are predominantly tropical in distribution, though they have a wide and fairly considerable representation in more temperate regions, notably in North America and South Africa.

These characters help to give to these plants a definite though not easily describable facies which tends to obscure the fact that they comprise two subgroups, usually treated as subfamilies, remarkably different in the details of their floral mechanisms. In both the pollen is removed *en masse* by insect agency, but in the smaller of them, the Periplocoideae, the pollen, in tetrads or masses of tetrads, falls naturally out of the anthers into or on to special structures which may be removed from the flower on the heads or bodies of visiting insects. In the larger subgroup, the Cynanchoideae, the stamens and styles are combined into a fused structure called a *gynostegium* (sometimes spelt *gynostemium*) and the pollen is aggregated into more definite pollinia. These pollinia become joined in pairs by structures which make it possible for them to be removed from the flowers on the legs or mouthparts of visiting insects. In neither case is the state of affairs exactly comparable with the organization in the monandrous orchids,* where the pollinia are removed in rather different fashion, and hence there are, in these three groups of plants, as many distinct variations within one and the same peculiar method of pollination.

Although their floral structure marks the Asclepiads off sharply from other Dicotyledons, and invites classification by which this is emphasized, they are, apart from this, closely similar to another group of flowering plants, the Apocynaceae, and this in a very interesting way. Indeed, although it is true that the Cynanchoideae are absolutely distinct in floral design, the Periplocoideae are not so, and between the simpler-flowered members of this latter group and some of the Apocynads there is scarcely any discontinuity, a fact which has been expressed by combining the two families into an order of their own, the Apocynales. But the Apocynads are, in

---

* See Summerhayes, V. S., *The Wild Orchids of Britain* (London, 1951).

their turn, divisible into two also, the subfamilies Plumerioideae and Echitoideae, and the differences between these are such that the former is a much more heterogeneous unit, and much less like the Asclepiads, than the latter. Thus there are, within the *two* families as usually defined, *four* series of plants of which the generalized characters and mutual resemblances are as follows:

Plumerioideae—stamens free or loosely connected with the apical parts of the styles: anther lobes full of pollen: pollen granular: no translators: fruit various, often a berry: seeds often hairless.

Echitoideae—stamens joined to the apical parts of the styles: anther lobes empty at the base: pollen granular: no translators: fruit follicular: seeds hairy.

Periplocoideae—stamens not actually joined to the apical parts of the styles (stigma-head): anther lobes full of pollen: pollen in tetrads which fall into or on to translators: fruit follicular: seeds hairy.

Cynanchoideae—stamens joined to the apical parts of the styles (stigma-head): anther lobes empty either above or below: pollen in pollinia which are united in pairs by yoke-like translators: fruit follicular: seeds hairy.

Without in any way attempting to suggest what the mutual phylogenetic relationships between these four may be it is clear enough that they represent four levels of particularization, or specialization, using these words to mean the gradual crystallization of one or more definite patterns of floral design. To summarize these, it may be said that the first shows the greatest heterogeneity and also that the apocynad characteristics are scarcely formulated: that in the second the characteristic fruits are general, while there is some degree of floral fusion, and partial emptiness of the anthers: that the third shows the simpler kind of asclepiad mechanism: and that in the fourth there is a full and peculiar pollinial condition, accompanied by partial emptiness of the anthers.

Pollination in the Asclepiads

1. Periplocoideae

The Periplocoideae are rather unfamiliar plants, and in temperate lands are known especially by the South European climbing

FIG. 99. *Periploca graeca:* habit, × ⅔, after
Bot. Mag.

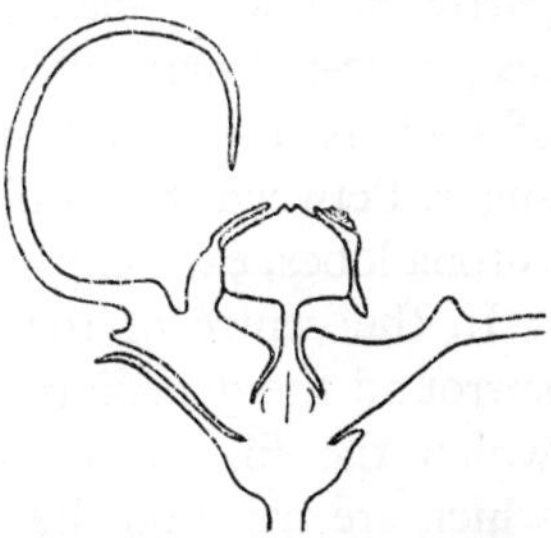

FIG. 102. *Periploca graeca:*
vertical section of flower,
× 3

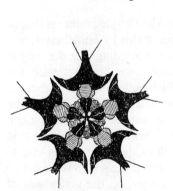

FIG. 101. *Periploca graeca:* centre
of an older flower, × 4

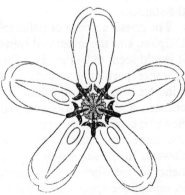

FIG. 100. *Periploca graeca:* plan of
flower, × 2, hairs omitted

species *Periploca graeca,* which is more or less hardy in this country, where it often flowers, and sometimes fruits, extremely well out of doors. The following account is taken from this species, which is illustrated in figs. 99–102.

The calyx lobes are small and only in the earliest stages cover the rest of the flower. At other times the outer green surface of the bud is the under, or abaxial, surface of the corolla lobes. These partly hairy lobes have a characteristic asymmetric shape and are deep rose-red madder with a green distal edge. Towards the base of each is a whitish patch of glandular papillae which secrete sugar. Between the corolla lobes are five incurved filiform maroon corona lobes, each expanded laterally at the base.

In the centre of the flower the five separate stamens closely surround the gynecium, to the mushroom-shaped stigma-head of which the filaments arc very closely appressed. The anthers, which are hairy on the back, are incurved over the stigma-head. There are two more or less completely superior carpels, free from one another below the stigma-head.

During the development of the flower material is secreted from the surface of the stigma-head at five points between the stamens which comes to form five translators, each somewhat the shape of a distorted wine-glass and consisting of three parts, a basal adhesive disc, a stalk, and an apical spoon or shovel. These translators are so situated that the spoon of each receives the pollen from one half of each of two adjacent anthers, while the adhesive discs hang down below the edge of the stigma-head between the filaments of the stamens.

The classic account of pollination in the Periplocoideae is that of Delpino, and the annotated translation of this by Hildebrand, and it would seem to be based on *Periploca.* According to it the tongue or other part of a visiting insect comes into contact with the adhesive base of the translator into which pollen has fallen from the anthers, and this sticks to it, so that when the insect leaves the flower it carries off the translator with its contained pollen in much the same way that the pollinia are carried off in many orchid flowers. In the same way too, the stalk of the translator is said to bend, so that when a second flower is visited the pollen is deposited in one of the spaces, between two stamens, on the under side of the stigma-head.

2. Cynanchoideae

The floral mechanism of the Cynanchoideae is a good deal more familiar, chiefly because it has been described often for *Asclepias*, and it is better therefore to deal here with the flower of a rather different and less well-known asclepiad, the beautiful Brazilian climber *Oxypetalum caeruleum*, which is of easy culture in greenhouses in this country. This is illustrated in figs. 103–105.

There are five small green sepals and five bluntly rounded oblong pale blue corolla lobes, the actual centre of the flower being shallowly campanulate. The corona lobes are upright and oblong, like the corolla lobes in colour (except that they are of a deeper tint where the tips are outrolled) and arise at their base, though alternating with them. The stigma-head is a long conical pointed structure to the base of which the stamens are adnate. The individual anthers, which are somewhat winged, have straight and parallel sides and are separated only by long narrow clefts within which are interstaminal chambers.

At the top of each anther cleft lies the chief part of a translator, the notched corpusculum (or retinaculum as it is sometimes called), which in the case of *Oxypetalum* is uncommonly long and conspicuous. At the base of it are the toothed translator arms, from each of which a pollinium hangs, one from each of two adjacent anthers. The translators arise in the same way as those of the Periplocoideae.

The usual account of pollination in the Cynanchoideae springs from the observations of Sprengel, or more accurately Robert Brown, since the latter was the first to realize the peculiarities of the asclepiad flower, but there are later descriptions by Delpino, Hildebrand and others. Most of these are based on the genus *Asclepias* and depict the process, in that genus at least, as follows. The flowers are visited by appropriate insects, notably bees and wasps, for their nectar. An insect crawling about over the flowers is sooner or later trapped through one of its legs becoming caught, usually at the tarsi, in one of the clefts between adjacent anthers. The insect can release itself only by drawing the limb upwards right through the cleft, and this it does, but as the leg becomes free at the top of the cleft it catches in the notch of the corpusculum, so that further movement pulls this, together with its attached

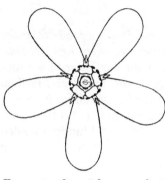

FIG. 104. *Oxypetalum caeruleum,*
plan of flower, × 1½

FIG. 103. *Oxypetalum caeruleum:*
habit, × ½

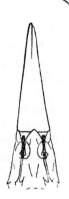

FIG. 105. *Oxypetalum caeruleum:* vertical section of flower,
× 5; gynostegium, × c. 8; corona lobe, × 8; pollinia, × 16

pollinia, away from its anchorage on the gynostegium. The released insect in due course visits another flower or inflorescence and again becomes caught by the leg in the same way. This time, however, during the drawing of the limb through the anther-cleft the pollinia are torn away from the corpusculum and left behind in the interstaminal chamber, and only the corpusculum remains attached to the insect's leg, where it offers no impediment to a repetition of the whole process. Most accounts add that as soon as the pollinia are pulled out of the anther the translator arms which unite them with the corpusculum contract in such a way as to bring the pollinia closer together so that they are the more easily deposited in the correct position near the stigmas of the next flower visited.

In the Cynanchoideae, though less so in the Periplocoideae, the flowers are of all sorts of shape and size, and the process of pollination is thought to take various expressions, particularly when the flowers are very tiny or very deeply tubular. In the latter case the usual explanation is that pollination is by long-tongued insects, such as butterflies, which transport the pollinia, not on their legs, but on their extended probosces, or by insects so small that they can pass down to the base of the corolla tube, though it would seem doubtful whether insects of this size could effectively transfer the pollinia.

From this digression into floral mechanisms we must now return to the question of the structural resemblances and differences between Apocynads and Asclepiads, a subject which reveals some extremely interesting points.

It was formerly supposed that the two families could be separated absolutely on the presence or absence of translators and of agglutinated pollen but later statements have brought a modification of this view, and to understand their significance it is necessary, first of all, to refer to certain Apocynads not directly concerned. In various members of that group, including the genus *Vinca*, which contains the most familiar European representative, the periwinkle, the receptive part of the style is not the apex but an area a little way below and often on the edge of an annular swelling surrounding the style which goes by the name of the *ring*. This ring is commonly described as secreting a viscous fluid with which the tongues of visiting insects come in contact so that the pollen thereafter adheres to them. In *Lochnera*, which may be regarded as a tropical version of *Vinca*, the pollen, according to Delpino, falls from the anther on to the upper, non-receptive, part of the ring in five heaps, though there is nothing like a translator to catch it. *Vinca* and *Lochnera*, it may be noted, both belong to the Plumerioideae. *Apocynum*, of which a very complete account has been given by Woodson, belongs to the Echitoideae, and is remarkable in two ways. One species, *A. androsaemifolium*, which is a well-known member of the North American flora, has long been called a "fly-trap" because of the frequency with which insects are fatally caught in parts of the flower, though there is some difference of opinion as to how this actually happens. Sometimes it is said that the insects' legs or tongues are caught between the anthers; sometimes that their tongues are caught between the nectaries and the carpels, whither they have been thrust in search of honey. The interesting point however, is not how this happens but that it should occur at all because it is a similar, though temporary, trapping of insect visitors which is the essential feature in the pollination story of the Cynanchoideae. The Australian Apocynad *Lyonsia straminea* is also stated to be a fly-trapper.

It is of considerable interest, in relation to these two plants, that some Asclepiads themselves have also been accused of trapping insects beyond the possibility of escape. Kerner (*Natural History of Plants*), for instance, describes how the very fragrant *Araujia albens* is, in its native South America, visited, without harm to themselves, by humble-bees, but that when it is grown in other lands it is often visited by great numbers of moths which become caught in the anther-clefts by their probosces and perish. In cultivation under glass in Britain *Araujia hortorum* often sets seed and honey-bees may be caught bearing the pollinia on their probosces.

A second species of *Apocynum* is remarkable for a different reason, and has been carefully studied by Demeter (who refers to it as *A. cannabinum*). He describes here the presence, in the flower, of five small plates (*teller*) which he states are formed from the viscous material secreted by the stigma. These plates or patches, though functionless, he regards as directly homologous with the simpler kinds of translators, similarly derived from secretions, in the Periplocoideae. Not only so but he considers, from a study of *Periploca*, that the more complex kind of translator characteristic of the Cynanchoideae can in turn be derived fairly simply from those of the Periplocoideae, different though they appear at first sight to be. Thus, if Demeter is right, *Apocynum* shows traces of a character diagnostic of the Periplocoideae and the difference between this group and the Cynanchoideae may be smaller than is generally reckoned. It is noteworthy here that Robert Brown, who was the first to separate the Asclepiads from the Apocynads, placed his genus *Cryptolepis* in the latter because he failed at the time to see the very small translators of the specimens he happened to examine. The correct attribution of *Cryptolepis* was made some years later by Falconer in a paper with the somewhat ambiguous title "On a reformed character of the genus *Cryptolepis*, Brown" which is also worth noting because, though published in 1845, it contains the sentence "Yet it is very evident that the plant described above has the whole of the accessory stigmatic apparatus of the Asclepiadeae . . . although in a less considerable degree of evolution . . ."

There is an interesting point about the pollen too. Here again the older view was that the Apocynads invariably have pollen in the granular form, and Schumann expresses his surprise that there is known to him only one exception, *Condylocarpus*, a genus not otherwise very reminiscent of the Periplocoideae. Woodson, however, maintains that in *Apocynum* and *Poacynum* the pollen tetrads never break up into grains, but how far these are properly developed tetrads is not made quite clear, and it is elsewhere stated that the pollen of practically all the species of *Apocynum* is largely abortive, which suggests that the tetrad condition may be something of an abnormality. Even so Woodson's observations are of great interest, and, incidentally, he also quotes Demeter as saying that this tetradenous pollen separates only with difficulty and usually remains in masses which may be transported as such by insects, though of course in the ordinary way.

Again in the anthers the Apocynads and Asclepiads show corresponding states. In the Plumerioideae and the Periplocoideae the anthers are full of pollen, but in the Echitoideae and Cynanchoideae they are partly empty. Furthermore (and this is very curious) in one large section of the Cynanchoideae it is, according to Schumann, the basal part which is full, but in the Echitoideae and most of the Cynanchoideae it is the apical part. Another interesting point is that in certain Apocynads, including *Apocynum*, the partition between the pollen sacs of each half anther is easily lost so that the anthers appear to have two loculi only, a condition characteristic of the Cynanchoideae.

Special reference has already been made to the style and stigma of *Vinca*, but in fact almost throughout the Apocynads the upper part of the style is developed into some kind of localized expanded structure, to which, in the case of the Echitoideae, the stamens are united, and so close and obvious is the parallel between this condition and that in the Asclepiads that Schumann used the term *narbenkopf* (stigma-head) for both families. In both it is some part of this stigma-head which is stigmatically receptive, and equally in both it is not uncommon to find the stigma-head surmounted by what appear to be ordinary apical stigmas or stigmatic branches but which are without receptive function. The occurrence of the "false stigmas" is one of the most puzzling minor problems of the Asclepiads and is discussed at greater length in the next chapter.

The corona is also not so complete a mark of distinction between the Apocynads and Asclepiads as might be expected, since, while it is almost invariably present in the latter, it is also not infrequent in the former, as for instance in *Melodinus, Diplorrhynchus, Rhazya, Vinca, Vallesia* and *Rauwolfia* among the Plumerioideae, and in *Adenium, Nerium, Strophanthus, Wrightia* and *Prestonia* among the Echitoideae. It is true that these apocynad coronas are generally in the throat of the corolla, while asclepiad coronas are much more commonly close to or actually upon the stamens, but these differences only add to their interest and probably some at least of each are homologous.

Rather similarly various Apocynads, including *Arduina, Allamanda, Tabernaemontana* and *Thevetia* in the Plumerioideae and *Echites, Nerium, Trachelospermum, Ichnocarpus, Strophanthus, Wrightia, Lyonsia, Parsonia* and *Prestonia* in the Echitoideae, have glands inside the calyx, a feature common in the Asclepiads.

On the other side of the account there is apparently at least one good general distinction between the Apocynads and the Asclepiads, although it is not absolute. In the former there is commonly, that is to say in a minority of Plumerioideae and in a great majority of Echitoideae a disc, which is annular or deeper and more or less lobed, and this is unrepresented in the latter. The Asclepiads have, however, as commonly, a corona closely associated with the stamens, a condition not exactly paralleled in the Apocynads. Thus, each has a particular floral item which, though in a sense the counterparts of one another, are not homologous, the asclepiad corona being outside the andrecium and the apocynad disc within it.

So much for the resemblances and differences between the Apocynads and Asclepiads, of which this somewhat lengthy review has served two purposes. First, it has shown the curious and interesting way in which what may be called border-line characters are distributed between the two, and how little in some directions in consequence there is any real distinction between them. Among the extremes of the two there is of course a great gulf but this is due more to an accumulation of condition or a concentration of characteristic in one of them rather than to any absolute difference in characters. At least traces of almost all asclepiad features can be found in apocynads, but only in the former do they achieve a distinctive unity of plan. Second, it has stressed the fact that the real distinction between the two is functional rather than structural. In other words the two groups comprise as a whole, partly plants in which a particular form of organization is in such a state of development as to result in a special kind of function, and partly plants in which this is not so and which, in consequence, function in a different manner. The taxonomist categorizes the former as the Asclepiadaceae and the latter as the Apocynaceae, and how the functional peculiarity of the former may have come about is the leading evolutionary problem concerning them.

The first point to be noted in the further examination of the Asclepiads, on which we shall henceforth concentrate, is the great disparity of size between the Periplocoideae, which are confined to the Old World, and the Cynanchoideae, which are widespread in both Old and New Worlds. The latter show a vastly wider range of form, and this is expressed in the fact that while in the former no more than forty genera, with less than 200 species, have been recognized, the latter have been divided into more than 250 genera and well over 3,000 species. Moreover, the Periplocoideae, though quite diverse for their numbers, are basically all much of a muchness, while the Cynanchoideae include several distinctive subsidiary types, the taxonomists' tribes, and the characterization of these is very interesting.

The state of affairs here can best be expressed briefly by using the modern conceptions of these tribes and arranging them in a dichotomous key, as follows:

Pollinia in the basal parts of the anthers
     ASCLEPIADEAE (figs. 103–109), American and African
Pollinia in the apical parts of the anthers
 Pollinia two in each anther lobe
     SECAMONEAE, African and Asiatic
 Pollinia single in each anther lobe
  Anthers dehiscing longitudinally
     TYLOPHOREAE (figs. 110–121), American, African
              and Asiatic
  Anthers dehiscing transversely
     GONOLOBEAE (figs. 122–125), American only.

The second and fourth of these tribes are not only much smaller in genera and species than the others but their distinctiveness does not compare with that between the other two, and it is therefore fair to regard the Cynanchoideae as comprising in the main two great assemblages, each with more than 1,500 species, in one of which (Asclepiadeae) the pollinia are in the basal parts of the anthers and hence normally pendent on the arms of the translators, and in the other of which (Tylophoreae, Secamoneae and Gonolobeae) the pollinia are in the apical parts of the anthers and hence normally erect on the translator arms, a distinction illustrated in figs. 105 and 111. Thus the whole subfamily divides into two on what can only be described as a minor, though nevertheless extremely clear-cut, character of the anthers. Moreover, as will be seen later, various other characters show a notable degree of correlation with this feature.

The tribe Asclepiadeae, rather like the smaller subfamily Periplocoideae, are so homogeneous that it is difficult to divide them into subtribes and the subject is not worth pursuing here, but the other large tribe, the Tylophoreae, comprises two easily distinguishable components. The formal recognition of these rests on characters of the stamens but correlated with this is a more generalized difference, one of them, the Ceropeginae, containing a notable proportion of herbaceous erect plants with inflorescences of few large flowers, and the other the Marsdeninae, being composed mostly of woody climbers with multiflorous inflorescences. The former again divide into two, the Ceropegieae, which are remarkable for their bizarre flowers, and the Stapelieae, which are succulent plants.

The presence in the Asclepiads of a corona showing an extraordinary range of form adds to the floral organization a variable ingredient which has been used extensively in the recognition of genera, though not altogether satisfactorily. This is particularly true of the Asclepiadeae, where the genera *Cynanchum*, *Asclepias*, *Gomphocarpus*, *Schizoglossum*, *Xysmalobium* and a set of smaller satellites, which are all separated from one another wholly or mainly on coronal characters, really make up what is a single plexus of species, most of which are erect plants and either herbaceous or at most suffruticose, and which provides almost all the members of the family in more temperate regions. Similarly in the Marsdeninae the pan-tropical genus *Marsdenia* is scarcely to be distinguished fundamentally from the genera *Hoya*, *Dischidia*, *Tylophora* and *Gymnema*, so that here again there is a huge plexus, in this case composed in the main of woody or at least subwoody climbing plants with many-flowered umbels of rather small flowers and characteristic of more tropical regions. Thus in the two great plexuses of the Cynanchoideae the erect herbaceous habit goes with apically sterile anthers and pendent pollinia and the more woody and climbing or twining habit with basally sterile anthers and erect pollinia.

Characters and their Distribution in the Periplocoideae

Since nearly all the features of this, the smaller, subfamily are seen more fully expressed in the Cynanchoideae, nothing but quite a brief summary of them is needed here.

The generalized periplocaceous plant is a rather slender, climbing or erect, woody plant, scarcely hairy, and with entire acute exstipulate opposite ovate-lanceolate leaves of moderate size and small flowers in multiflorous racemose cymes which are either terminal or truly axillary. The corolla is rotate and has a simple corona near the stamens, which have apical appendages. The stigma-head is flattened and bears spoon-like translators, and the smooth follicular fruits contain many flattened seeds with silky hair-tufts.

From this norm departure is nowhere very great, though most characters show some range of variation, and the following is an epitome of the more egregious states.

Habit. Many are epiphytes. *Utleria* is described as a small tree looking much like an Apocynad. *Raphionacme* shows most specialization in habit, some species having herbaceous stems rising from much-swollen underground stocks. *Periploca* includes species which are more or less leafless and switch-like or somewhat fleshy. Hairiness is seldom copious.

Leaves. Apart from variation in relative width, the most notable point is venation, in which the pinnate laterals may be very numerous and parallel. The leaves are sometimes in whorls or clustered at the ends of the branches: in *Utleria* they are alternate.

Inflorescences. The number of flowers varies greatly, as also does the length of peduncles and pedicels. *Telectadium* is said to have a scorpioid cyme.

Flowers. The size varies from $\frac{1}{10}$th inch diameter in *Utleria* to $2\frac{1}{2}$ inches in *Raphionacme*. Colour is also variable, some of the more unusual tints, such as browns, greens and maroon are notable, and there are some striking combinations.

Corolla. The corolla is nearly always rotate or campanulate-rotate, but in at least three cases is hypocrateriform. In *Phyllanthera* it is urceolate, and in *Telectadium* it has a narrow cylindrical tube and a dilated basal part. The degree of lobing varies and the lobes may be completely reflexed. There is a peculiar distribution of hairs on the corolla of *Finlaysonia*.

Corona. This is altogether absent in four genera and in the rest varies considerably in position, a distinction commonly being drawn between those which are in or near the sinuses of the corolla and which may therefore be far removed from the gynostegium, and those which are close to or even adnate to the stamens. It is the more distant sinusal coronas of *Cryptolepis*, *Cryptostegia*, *Ectadium*, *Pentopetia* (*Petopentia*) and a few others which are presumably homologous with the coronas in Apocynads. The corona is double in *Omphalogonus*. The shape of the lobes is very variable.

Stamens. These vary chiefly in the degree to which the appendages, which appear to be homologous with the "end-bodies" of Schumann in the Cynanchoideae, are developed. They have the form of long hair tails in *Ectadium, Pentanura* and *Pentopetia.*

Pollen. Apparently always in tetrads but these may themselves be quite free from one another, as in *Cryptolepis,* or may cohere so that they fall from the anther in loosely aggregated masses, which come to pieces when touched. These masses may be two *per* anther lobe or even more in *Gymnanthera.*

Translators. These vary considerably in design, and since they provide the ultimate characterization of the Periplocoideae deserves rather more notice here. They are formed from a secretion from parts of the stigma-head and are therefore structures of a very peculiar kind. They range in form from flat, narrow, ligulate strips, as in *Cryptolepis,* to well-defined more or less erect spoon-like or even goblet-shaped structures. They nearly always consist of spoon, stalk and disc. Most commonly they are referred to as spoon-, ladle- or shovel-shaped, while intermediate between these and the flat forms are the so-called gutter-shaped kinds of *Telectadium* and *Chlorocodon.* In the flatter sorts the pollen adheres by the viscosity of the surface of the translator, but in deeper and more cup-shaped kinds adhesion is not so important, and it seems that it is in association with such deeper translators that the larger and more coherent pollen masses occur. Deep translators are well seen in *Gymnolaima* and *Phyllanthera,* where they have been compared with cornucopias, and in *Hemidesmus* and *Camptocarpus,* where they are described as first cup-shaped and later flattened.

Fruits and seeds. The fruits of Periplocoideae show comparatively little variation but in *Finlaysonia* the seeds are without hair-tufts. The genus *Finlaysonia,* it may be noted, is also anomalous in general appearance, particularly in the thick leathery leaves and in the stout inflorescence axes, and is also unusual in living in marine mud.

Characters and their Distribution in the Cynanchoideae

It has been said already that in terms of species numbers the Cynanchoideae are fifteen times as large a group as the Periplocoideae and it may be emphasized here that they differ also, in two respects, in their constitution. One of these respects, already noted, is that the larger group comprises several well-marked subtypes, but a more important difference is that the greater species numbers of the Cynanchoideae are largely accounted for by the development of a number of large and close species plexuses, or, in taxonomic phrase, of large genera with closely similar species. Thus while the Periplocoideae consists of comparatively few and relatively small genera more or less equally distinct from one another the Cynanchoideae consist partly of such genera but much more of large genera less isolated from one another, with the effect that while it is difficult to find any predominating type in the Periplocoideae, there is a small number of

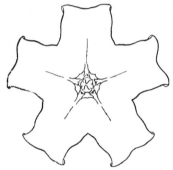

FIG. 106. *Araujia hortorum:* plan of flower, × 2

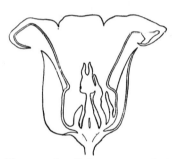

FIG. 107. *Araujia hortorum:* vertical section of flower, × 2

very strongly predominating types in the Cynanchoideae. In short, in this latter group, certain designs are expressed far more elaborately and multifariously than in the other. Thus, it is clear that the range and variety of structure in the Cynanchoideae are immensely greater than in the Periplocoideae and will require much more lengthy description and consideration. Unfortunately, space alone forbids an exhaustive treatment and it is inevitable therefore that many minor facts of great interest will have to be ignored. Where no other group is specifically mentioned all references are to the Cynanchoideae as a whole.

Habit. About half the Cynanchoideae are climbing shrubs, but species range from robust woody climbers properly called lianes to such small plants as the one annual herb, *Conomitra,* or as *Steinheilia radicans,* which is a perennial herb, with a stem of only an inch or two arising from a swollen underground stock. Points of general interest in habit are five:

1. The tribes and subtribes vary greatly in range of habit. The small Secamoneae and Gonolobeae are almost exclusively woody climbers. The Asclepiadeae include a great number of leafy herbaceous forms. In the Tylophoreae the two subtribes are very different. The Marsdeninae are nearly all climbers, mostly woody, and among them are many epiphytes with adventitious roots. The Ceropeginae consist very largely either of tuberous herbaceous forms or of cactoid plants.

2. The great development of herbaceous forms is in the Asclepiadeae and in the Ceropeginae but while the former contain many erect perennial herbs the herbaceous forms of the latter are mostly either decumbent or scrambling herbs from strongly tuberous bases or succulents. The erect and sometimes almost fastigiate types of Asclepiadeae are associated with more temperate latitudes, and the tuberous and succulent types of the Ceropeginae with the drier parts of warmer places. It should be noted that all the larger genera of Asclepiadeae are more or less herbaceous, *Asclepias* and its consorts having erect stems and *Cynanchum* having more or less climbing stems.

3. Reduction or entire absence of leaf occurs widely but unevenly, and is usually accompanied by some degree of succulence in the stems. *Decanema* and *Decanemopsis* are described as leafless lianes. Several of the Asclepiadeae are leafless, at least when mature, among them *Absolmsia spartioides*, which has somewhat fleshy branches. In the Ceropeginae fleshy plants are outstanding in the Stapelieae and a few species of *Ceropegia* have leafless fleshy stems, e.g. *C. stapeliaeformis* from the Cape and *C. fusca* from the Canaries. There are leafless switch plants in *Leptadenia* and *Orthanthera*.

The Stapelieae consist, with one exception, of succulent cactoid plants, the exception being the monotypic genus *Frerea*, which is of special interest because it is permanently leafy, and thus bears to the rest of its group the relation that the genera *Pereskia* and *Maiheunia* bear to the rest of the Cacti. In all but this genus the stems are short and cylindrical or polygonal in section, or even subspherical, and, at least at maturity, devoid of leaves, which are replaced by spines or hairs. Nearly all are smaller and more low-growing than most of the Cacti or succulent Spurges, but to similar-sized members of these groups they bear a very close superficial resemblance.

4. The Ceropegieae, dominated by the two great genera *Ceropegia* and *Brachystelma*, between which the separation is but formal, show a remarkable range of habit best illustrated by a few examples. Some species of *Ceropegia* are leafy, more or less woody, climbers: others have long green trailing subfleshy stems with thick leaves: some have trailing wiry stems bearing tubers: many others have variously small herbaceous stems arising from swollen

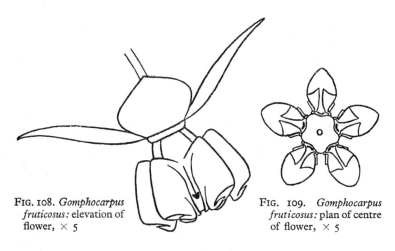

FIG. 108. *Gomphocarpus fruticosus:* elevation of flower, × 5

FIG. 109. *Gomphocarpus fruticosus:* plan of centre of flower, × 5

underground stocks: while a few species are succulent. In *Brachystelma* and *Decaceras*, also, there are many with herbaceous stems from subterranean or half-buried stocks, and the corm-like tubers of *B. barberiae* are very striking. This form is perhaps the most prevalent in the group. Sometimes, as in *Macropetalum*, the stems are very delicate. Finally in *Leptadenia* some species are woody climbers and others are leafless "ephedroid" switch plants.

5. There are frequent instances of superficial resemblance between members of different genera in different parts of the world, and Schumann refers to a number of these. This effect may be the result of false classification but even if this is so it at least indicates that there is little correlation between habit and the floral structure on which formal classification is more directly

based. As examples of this resemblance there may be mentioned that *Glossonema angolense* is said to be like members of *Asclepias* sect. *Pachycarpus* except in the corona; and that *Woodia mucronata* var. *bifurcata*, *Parapodium crispum* and *Xysmalobium gomphocarpoides* are said easily to be mistaken for one another. The number of specific epithets which are based on generic names suggests that there are many other such cases.

Leaf. The main theme of variation in leaf form is in size (though there is also some range of fleshiness), culminating in one direction in almost complete leaflessness, and apart from this, which needs no further comment here, there are only a few minor, and one outstanding, deviations. The former comprise the lobing of the leaves which is seen in *Tapeinosperma*, where they are described as irregularly toothed ; in *Eustegia* where they are more or less three-lobed; and in the aberrant genus *Emicocarpus* where they are deeply five- to seven-lobed. The latter is that of the remarkable leaf modifications which are to be seen in species of the closely related genera *Dischidia*, *Conchophyllum* and *Hoya*, and which is best known in the so-called "pitcher-leaves" of the first-named. Among these plants there are three peculiar conditions, namely, the concave or shell-like leaf of *Conchophyllum* and of certain species of *Hoya*, which is appressed like a shield to the trunk of the tree over which the plant is growing: the simple pitcher-leaf of *Dischidia*, best known in *D. rafflesiana*; and the double pitcher-leaf of certain other members of the genus, among them *D. complex* and *D. pectenoides*.

Corolla. The Cynanchoideae exhibit, in their corollas, elaboration of a kind only surpassed in that of the corona, and the flowers are therefore very various in form and appearance. Nowhere, however, is this extraordinary range due to any real zygomorphy, and except for some slight positional obliquity in the outer flowers of some inflorescences and for curvature of the tube in some deeply tubular corollas, chiefly in species of *Ceropegia*, the flowers are invariably regular. On the other hand variety is accentuated by a remarkable gamut of flower colour and combinations of colours, many of an unusual kind.

A useful approach to this complicated matter of flower form is the realization that the different kinds can without very much difficulty be recognized as conditions in the variation gamut of a single characterization namely the degree to which the corolla opens and exposes the reproductive parts of the flower at anthesis, and it will help to explain this by saying at once that at one end of this gamut there are corollas strongly urceolate and with the smallest of openings (figs. 114, 115) while at the other are corollas which early become so completely reflexed that the gynostegium

FIG. 110. *Hoya carnosa:* plan of flower, × 4, hairs omitted

FIG. 111. *Hoya carnosa:* vertical section of flower, × 4; gynostegium, × 6; pollinia, × 22

is the most conspicuous part of the flower (fig. 108). If it be remembered also that throughout this gamut there may coincidently be almost any degree of gamopetaly between the petals and almost any range of flower size, the bewildering display of floral form in the group may at least be seen to possess some fundamental order.

Although there is this great range of corolla form many of the contributory types are uncommon or rare, and indeed many of the more extreme forms are confined to relatively few genera.

The difficulties of making numerical assessments of this kind are manifest especially when it is remembered that some generic distinctions ignore this character but it is worth pointing out that rotate or rotate-campanulate corollas are much the commonest and may be said to represent the basic generalized form: that campanulate corollas come next in number: and that urceolate or tubular corollas are comparatively few. It is noteworthy that nearly all the largest genera exhibit the first of these.

As has already been said analysis of corolla form in the Cynanchoideae can be conveniently based on the degree of reflexing of the corolla and the consequent extent to which the parts are exposed at anthesis, and on the degree of gamopetaly, and to these may be added the proportion of the corolla which is reflexed and the shapes of its free lobes. Flower size, though in one sense contributing more than anything else to appearance, is best ignored as quantitative rather than qualitative.

These four then—degree of exposure, degree of gamopetaly, proportion of corolla reflexed, and the shape of the corolla lobes—in various combination give the main corolla plans and would account for most of the more ordinary states, but there are in addition a number of other conditions which, more or less unusual in themselves, produce marked peculiarity when present to any considerable degree or in any strong combination. These include the folding of the corolla lobes lengthwise; the coiling or twisting of the lobes; the curving of the corolla tube; various elaborations of the teeth or lobes; fusing of the tips of the lobes (or their failure to separate as the flower opens); and the development of a raised annular ring or cushion below the sinuses. There are also flower colour and hairiness.

The flower colour of the Cynanchoideae is, at least to European eyes, one of their most conspicuous features. Greens, browns, and the more lurid purples, often almost black, occur frequently in combination, as do whites, paler yellows and paler purples. Strong yellows and reds seem to be less frequent though there is a good example in one of the most familiar members of the group, *Asclepias curassavica*, and blue is perhaps the rarest, though here again is the well-known example of *Oxypetalum caeruleum*. Contrasts are often striking, as in *Ceropegia*, where the colours are often zoned, and in many Stapelieae. Counterveining, freckling and striping are all frequent, again particularly among the Stapelieae and in *Ceropegia*. In some others, notably the *Asclepias* plexus and the genus *Hoya*, the corollas are plain-coloured, but very diverse effects result because the coronas may be either of the same colour or of a contrasting one. In *Hoya*, again, the corolla is not uncommonly white while the corona is strongly coloured and appears as a smaller star against the white background, giving a "flower-in-flower" effect. This great range of colour is in striking contrast with certain other families, but it is of interest to note how generally similar it is to the range in the Orchids, the other great pollinial group.

Ornamentation in the form of hairs and glands contributes to

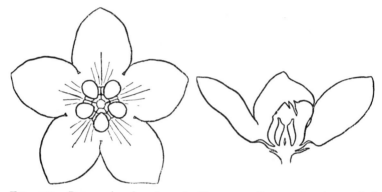

FIG. 112. *Dregea sinensis:* plan of   FIG. 113. *Dregea sinensis:* vertical
    flower, × 4   section of flower, × 4

some of the most striking floral effects in the Cynanchoideae but the subject is a difficult one to deal with here and the reader should refer to pictures of the family and particularly to the beautifully illustrated monograph of the Stapelieae by White and Sloane to gain some general impression of its importance, though a few comments can be made. First, the development of hairs, though widespread, is very uneven in the group, being most conspicuous in the Marsdeninae and the Stapelieae and least so perhaps in the Asclepiadeae, but extremely hairy forms may occur here and there throughout. Within the individual flower the hairiness may be general and to an extent which almost conceals the corolla as in *Macroscepis* and some species of *Hoya*, or restricted to tufts or fringes, or be distributed in a definite pattern as in many species of *Marsdenia*. In *Trichosacme* the corolla lobes are glabrous but bear long tail-like appendages which are densely hairy. More extensive flat glands sometimes occur on the corolla as in *Decaceras* and *Leptadenia*, and a most extraordinary example of this is to be seen in *Barjonia cymosa*. Here there is a small sessile reniform gland about the middle of each lobe, and this is connected with the base of the lobe by a median line of hairs, so that the effect is that of a normal stamen with a reniform anther and hairy filament lying against the corolla lobe. In the Stapelieae the extreme conditions are too numerous to list but there may be mentioned as examples the great cushion of pink hairs which fills the centre of the brown and yellow flower of *Stapelia pulvinata*; the soft spine-like processes that cover the corolla of *Huernia hystrix*; the curiously swollen hairs of *Caralluma stalagmifera*; the curious papillate spines of *Pectinaria asperifolia*; the hairy halo round the corolla ring of *Duvalia corderoyi*; and the strange vibratile deciduous or caducous hairs in varous species of *Stapelia* and *Caralluma*.

Lastly with regard to the corolla is its size. The flowers of *Gymnema glabrum* are described as from $\frac{1}{12}$th to $\frac{1}{16}$th inch across and those of *Dischidia imbricata* are no more than 1·5 mm. long, and from this extreme there is every size up to the enormous flowers of *Stapelia gigantea*, which may be 16 inches across. At the same time nearly all the larger flowers, say above 3 inches across, are in the Stapelieae, and in other groups the average size is less than 1 inch. In total most of the Cynanchoideae would seem to have flowers of about $\frac{1}{2}$ an inch across. The smallest are chiefly in the Marsdeninae. This great range of size is remarkable enough in itself, and in sufficiently striking contrast to some other families, but it is particularly noteworthy in plants with insect pollination of so precise a kind, because, although there is no direct and constant relation between the size of the gynostegium and that of the corolla both vary in this respect, and consequently in the sort of insect visitors appropriate.

Corona. The almost invariable presence of a corona, that is to say of a structure or series of structures "intervening between the corolla and andrecium" is certainly one of the most remarkable features of the Asclepiad flower, and in the Cynanchoideae it is seen in more bewildering variety than in any other group of plants.

FIG. 114. *Dischidia albida:* elevation of flower, × 6

FIG. 115. *Dischidia albida:* vertical section of flower, × 6

FIG. 116. *Marsdenia oreopila:* plan of flower, × 4

FIG. 117. *Marsdenia oreopila:* pollinia, × 12; vertical section of flower, × 4

So much so that it is quite impossible here to do more than touch upon some of the outstanding aspects of the subject. First and foremost of these is the astonishing demonstration that these coronas afford of what is commonly called *elaborative evolution,* for they are found almost throughout the group and in almost every condition between several widely divergent extremes. At the same time there is an obvious degree of instability about them and this makes immoderate use of them for taxonomic purposes both tiresome and undesirable. There are, for instance, genera containing species with very different coronas; genera restricted each to one coronal form; intermediate coronal forms between genera; varieties of species distinguished by coronal differences; and even, on occasion, more than one form of corona, or at least a difference of coronal shape, on one and the same individual plant. In considering the corona it is therefore specially important to let the facts speak for themselves and to rid one's mind of any preconceived ideas as to the relation between coronal form and other floral characters and as to the importance of the former in indicating classificatory status or phylogenetic relation. It may be added that while some, and possibly many, coronas are nectarial, as is obvious from the glistening beads of honey secreted by *Hoya* flowers or the accumulation of sugary liquid in the bucket-like corona lobes of *Asclepias,* others reveal no particular role in the biology of the flower, and even seem, in some cases, to be a hindrance to pollination rather than a help, though this is of course a purely human estimate.

Various opinions have been expressed as to the origin and homologies of the Asclepiad corona, but it must suffice here to remind the reader that in the Cynanchoideae, as in the Periplocoideae, it is customary to recognize two sorts of coronas according to their positions in respect of the corolla and gynecium, namely the *corolline* or sinusal, in which the corona lobes are situated at or near the sinuses of the corolla and therefore, unless the corolla is virtually polypetalous, at some distance from the gynostegium, and the *staminal,* in which the corona is inserted close to or actually on the stamens. It has been suggested that the coronas of the Periplocoideae are invariably of the former sort in origin and those of the Cynanchoideae invariably of the latter sort, but this would seem to be something of an oversimplification, and there are certainly corresponding appearances in the two subfamilies.

Corolline coronas, which are relatively infrequent, usually have the form of five separate scales, lobes or ridges, while staminal coronas are usually of five or ten free or more or less united members, sometimes apparently in two concentric rings. In the latter case the corona is often called double, but this is not always an accurate description. Some certainly are double in the true sense of being in, and forming, two distinct whorls, but most so called only appear to be double because the lobes are divaricately cleft or because they bear appendages on the inner side.

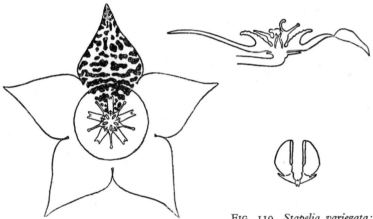

FIG. 118. *Stapelia variegata:* plan of flower, about natural size; the patterning is shown in one segment only.

FIG. 119. *Stapelia variegata:* vertical section of flower, about natural size; (below) pollinia, × 10

It is, unfortunately, quite impossible to give anything like an adequate account of the corona in the Cynanchoideae here, and the best that can be done in the space available, in order to stress the points of greatest evolutionary interest is to arrange the most prevalent or otherwise noteworthy types in the form of a brief, annotated, key. Where, in this key, the name of a genus is followed by an asterisk it means that the coronal type concerned occurs particularly in many species of that genus or plexus. Where the name is followed by "etc." this indicates that the unit mentioned affords one of the best, but not the only, examples. If a type appears to be rare or unique that is stated. It is also important to remember that these types are not all equally unlike and isolated from one another. True, some stand markedly apart from the rest but nearly all of them can be linked up in one direction or another by intermediate conditions.

Key representation of the coronal forms of the Cynanchoideae

A Corona absent   all or some species of 12 genera in the Asclepiadeae and 12 in the Tylophoreae

AA Corona present

  B Corona forming a single whorl or single ring

    C Corona corolline (1) Chiefly in a few small genera of Asclepiadeae

  CC Corona staminal

    D Corona clearly adnate to the filaments or anthers

        1 Corona of five small members (2)
          Marsdenia*, Tylophora*

        2 Corona of five horizontally compressed members (3)
          Gonolobus etc.

        3 Corona of five members similar to the anthers but usually smaller (4)
          Marsdenia*

        4 Corona of five radial keels (5)
          Secamone*

  DD Corona arising at or near the base of the filaments

    E Lobes of the corona more or less free

      F Corona lobes without appendages

        1 Corona of five small members (6)
          infrequent

        2 Corona of five more or less erect flat members (7)
          rather frequent in Asclepiadeae and Gonolobeae, *Metastelma* and *Oxypetalum* etc.

        3 Corona of five more or less horizontal members
          rare   [(8)

        4 Corona of five concave or cucullate members (9)
          Gomphocarpus*

        5 Corona of five convex members forming a flat or conical star (10)
          Hoya*

        6 Corona of five pitcher-like members (11)
          Stathmostelma etc.

        7 Corona lobes with coiled basal spurs (12)
          Calotropis only

        8 Corona lobes more or less divided (13)
          frequent

FF Corona lobes with appendages

  1 Corona lobes flat, appendages terminal (14)
   frequent
  2 Corona lobes flat, appendages usually small but
   not terminal (15)
    Oxypetalum etc.
  3 Corona lobes flat with terminal and other
   appendages (16)
   infrequent
  4 Corona lobes cucullate with an inner horn (17)
   Asclepias★

EE Corona lobes more or less joined at the base
 G Corona lobes not more than five

  1 Corona annular or shallowly cupulate (18)
   rare
  2 Corona a horizontal star (19)
   rare
  3 Corona toothed or lobed cucullate (20)
   Cynanchum★
  4 Corona lobes simple erect (21)
   various
  5 Corona lobes double hooked (22)
   Dischidia★
  6 Corona urceolate or tubular (23)
   Cynanchum etc.

 GG Corona lobes apparently more than five

  1 Ten similar corona lobes in one ring (24)
   rare
  2 Ten corona lobes of two sorts in one ring (25)
   infrequent
  3 More than ten corona lobes in one ring (26)
   rare, chiefly in Gonolobeae

BB Corona clearly or apparently in two or three whorls (double or
triple)
 H Corona of both corolline and staminal whorls (27)
    Marsdenia spp., *Secamone* etc.
 HH Corona not including a corolline whorl
  K Corona staminal with a small additional annular ring (28)
   rare
  KK Corona staminal without additional ring
   L Corona lobes apparently in two whorls
    M Outer and inner whorls both well developed
     N Outer lobes more or less free
      O Outer lobes more or less entire

P Outer lobes tapering or pointed

   1 Inner lobes simple (29)
     Stapelia, Caralluma etc.

   2 Inner lobes with dorsal wing or process (30)
     *Stapelia**

   3 Inner lobes with two clavate horns (31)
     Stapelia clavicorona only

   4 Outer lobes toothed at the base (32)
     Riocreuxia

PP Outer lobes abruptly truncate

   1 Outer lobes oblong more or less entire, inner simple (33)
     Huernia spp.

   2 Outer lobes oblong more or less entire, inner branched or keeled (34)
     *Stapelia**

   3 Inner lobes with one or both branches ending in tuberculate knobs (35)
     Stapelia spp.

   4 Outer lobes obcuneate (36)
     Caralluma spp.

OO Outer lobes more or less deeply divided

   1 Outer lobes ascending (37)
     Caralluma, Trichocaulon*

   2 Outer lobes horizontal (38)
     Trichocaulon etc.

   3 Branches of all lobes ending in tuberculate knobs (39)
     Stapelia longii

NN Outer lobes more or less joined
  Q Outer lobes forming a horizontal rounded plate

   1 Inner lobes more or less erect (40)
     Huernia spp.

   2 Inner lobes bent, dorsally swollen or keeled (41)
     *Duvalia**

QQ Outer whorl more or less cupulate
  R Inner whorl more or less similar to the outer (42)
     various
RR Inner and outer whorls markedly different

   1 Outer a plain or sinuous cupule (43)
     Caralluma, Ceropegia**

   2 Outer a cupule with plain bifid teeth (44)
     Caralluma, Ceropegia**

        3 Outer whorl with knobbed bifid teeth (45)
           Tavaresia and a variety of a *Caralluma*
           species
        4 Outer whorl forming a deep tube (46)
           Stapeliopsis only

  MM Outer and inner whorls not equally developed
    S Outer lobes much larger than the inner (47)
        Stapelianthus, *Podostigma*
   SS Outer lobes much smaller than the inner (48)
        Pectinaria etc.
 LL Corona lobes apparently in three whorls (49)
        Emicocarpus, *Eustegia*, *Neoschumannia* only

The outstanding points of evolutionary interest which emerge from the foregoing analysis of coronal form in the Cynanchoideae may be summarized as follows:

1. To a large degree the great variety of coronal form results from the occurrence of certain common character-components found widely in the plant kingdom, such as variation in absolute dimensions (diminution and augmentation), variation from free units to joined (coalescence and adnation), and variation from simple to divided or multiplied (fission and duplication). The last includes most of the accurately named double coronas.

2. This basic variation in coronal form is greatly enhanced by certain less general and more particular variations of which only two need be mentioned here, namely the production of appendages on the corona lobes, especially on the inner side, and the bending of the divisions of the corona inwards and outwards alternately. One or other or both of these account for the rest of the double coronas.

3. The colour and ornamentation of the corona lobes reveals both common and uncommon aspects. In general the range of colour is not very wide and conforms with that of the corolla but there are some striking exceptions to this, one of the most remarkable being the case of *Margaretta* where the larger corona lobes may be in strong contrast to the relatively small corolla lobes. The somewhat similar instances in *Asclepias* and *Hoya* which produce a "flower-in-flower" effect have already been referred to. Speckling and similar markings are not uncommon especially, as with the corolla, in the Stapelieae. The corona may, as in species of *Ceropegia*, be strongly coloured even though at the base of a deep

and narrow corolla tube. Finally, the corona may be hairy, as in *Hickenia scalae*, but this is a distinctly uncommon condition, and much more so than in the corolla.

4. Some coronal forms can only be described as exuberant in the sense that they have a kind and complexity of form which can scarcely be accounted for by the association of any special function. *Dictyanthus, Callaeolepium* and *Neoschumannia* are notable examples of this but it is again in the Stapelieae that the best instances are found, reaching an extreme in *Stapelia longii* and *S. gerstneri*.

5. Certain types of coronal form are characteristic of large and species-rich genera and are therefore, as might be expected, expressed in long series of rather similar states, differing, it may be, in no more than colour or *minutiae* of design, as may particularly be seen in *Marsdenia, Dischidia* and *Ceropegia*, and in such there are more or less complete series of intermediates between opposite extremes. On the other hand some coronal forms are quite isolated from all the rest and may be expressed in only a single species, which in consequence, is usually treated as constituting a separate monotypic genus. Some types seem to result from the effects of other characters, as for example in the horizontally compressed coronas of the Gonolobeae, in which the gynostegium is telescoped vertically.

6. There is considerable difference in the development of the corona in the various groups of the Cynanchoideae. In general it is least in the Secamoneae, though the Asclepiadeae also show a wide range of relatively simple forms. In the Marsdeninae it is, though more particularized, also of relatively simple design. In the Ceropeginae, on the other hand, the coronas are nearly all elaborate double types, while the Stapelieae have the most complex forms of all. The great mixture of coronal forms in the Gonolobeae is one of the reasons for the opinion that this group is not so natural as the rest, but *Gonolobus* and its true associates have a highly specialized type.

7. It has already been suggested that the corona has no obvious and indisputable function, and indeed its very multifariousness may be held to support this view, but there is one aspect of this question which is of exceptional interest, namely the fact that in some groups the corona assumes a form which makes the flower as a whole appear to have an organization which it does not

actually possess. Thus in *Margaretta*, as already mentioned, the corona lobes may be comparatively large and brightly coloured petaloid bodies which overshadow the smaller true petals in attractiveness, but the most remarkable example is that of some of the Stapelieae. In these plants, as in the rest of the Cynanchoideae, there is no trace of normal stamens, styles, and stigmas, such as in more ordinary flowers commonly form a centre of two five-rayed

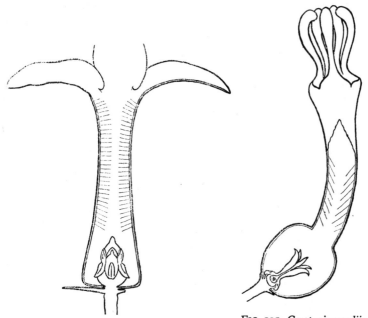

FIG. 120. *Stephanotis floribunda:* vertical section of flower, about twice natural size

FIG. 121. *Ceropegia woodii:* flower, partly in vertical section, × 3

whorls, but in some Stapelieae the double form of the corona exactly reproduces such a double whorl, the outer ring of lobes simulating stamens and the inner simulating styles (see particularly figs. 511, 514, 542, 551, 574, 604 in White and Sloane's monograph).

Gynostegium. This is the compound structure resulting from the fusion of the stamens and gynecium, and there are two points concerning it as a whole which call for notice before its component parts are dealt with. The first is the degree to which it, effectively, projects from the flower when the latter is fully open. In this there is considerable variation in which one extreme is the condition most commonly seen in the Gonolobeae, where the structure is so depressed as to appear almost telescoped, and does not project from the flower at all. Towards the other extreme is the *Asclepias* condition in which the gynostegium may be said to project fully from the flower, this projection being emphasized by the reflexing of the corolla. The real extreme in this direction is, however, provided by the few cases in which the tube formed by the fused staminal filaments is so long that the whole effective unit of corona lobes, stamens and stigma-head is raised well above the level of the remainder of the flower. In this state the stigma-head is often called stipitate and good examples include *Fanninia calostigma*, *Podostigma pubescens* and *Drakebrockmania crassa*. The second point is of interest chiefly as illustrating the possible origin of a characteristic in consequence of evolutionary change elsewhere. In the Cynanchoideae the stamens are fused to the stigma-head, which is itself a prolongation upwards of the carpels, and hence the ovary, the stigma-head, the stamens and the corolla (to which the stamens are attached) are all in organic continuity, so that ordinary corolla and staminal fall is impossible. Instead, a point of severance develops at the narrowest point of the carpel necks immediately below the stigma-head so that, in due course, the stigma-head and the corolla to which it is attached by way of the stamens fall off as a whole leaving the fertile carpels to expand without hindrance. This can be seen well in *Stephanotis*. One final small point about the gynostegium as a whole is that in *Dictyanthus*, which is in many respects an egregious type, it possesses an unusual degree of complexity.

Stamens. In the Cynanchoideae the stamens are a much more integral part of the pollination mechanism than they are in the Periplocoideae. They are always fused with the stigma-head at some point and the anthers are commonly more or less winged or curtained laterally so that the edges of adjacent anthers are so closely parallel that they form the clefts which guide the leg or proboscis of an insect towards the corpusculum. Below these clefts are spaces, of varying size and shape according to the form of the stamens and the stigma-head, called the interstaminal chambers, in which the pollinia torn from the insects' legs become deposited.

The filaments, if present, are fused together into what thus becomes the outer tube of the gynostegium. The actual structural range of the stamens is, however, very difficult to determine, chiefly because of the inadequacy of descriptions, and it would be of little profit to attempt a review of it here. It is better rather to concentrate on a few matters of special significance from an evolutionary point of view.

1. It seems clear that, throughout the Cynanchoideae, the anther is composed of two fundamental parts only, a basal part

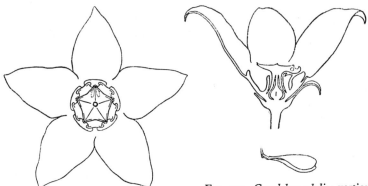

FIG. 122. *Gonolobus edulis:* plan of flower, × 3

FIG. 123. *Gonolobus edulis:* vertical section of flower, × 3; pollinia, × 12

(*grundkörper*) which incorporates the clefts, and an apical part (*endkörper*). The terminology is confusing but it would seem that these apices are, in origin, connective appendages, and homologous with the structures so called in the Periplocoideae. The all-important fact is that the pollinia may be either in the bases or in the apices of the anthers, that is to say if the interpretation suggested is correct, either in the anther proper or in part or all of its appendage. The former condition diagnoses the Asclepiadeae. In this group the pollinia are generally pendent while in the opposite condition they are mostly erect on the translators, because the position of the corpusculum relative to the stamen as a whole remains the same in both.

The method of dehiscence of the anthers is also associated with this difference. When the pollinia are pendent and pulled upwards out of the anther it is sufficient for the latter to dehisce by a comparatively restricted apical opening, but when the pollinia are erect on the translator-arms there must be longitudinal dehiscence of the anther in order to liberate them easily. Similarly when the pollinia are set horizontally the anther dehiscence is transverse.

2. The translator-unit, namely the corpusculum, the translator-arms and the pollinia, shows considerable variation, especially in its total size, in the shape of the pollinia proper, in the shape and length of the translator-arms, and in the shape and size of the corpusculum. As regards the first point the largest units are apparently those of *Stathmostelma*, which may be nearly a quarter of an inch wide and the same in height. In many related plants, such as *Asclepias*, the height is about $\frac{1}{16}$th of an inch and from here there is every smaller state to such genera as *Fockea* and *Henrya* where this dimension is no more than $\frac{1}{150}$th of an inch. A fair average seems to be about $\frac{1}{50}$th of an inch in one or other direction.

3. The pollinia are generally clavate if they are pendent, and oval, ovate or discoid if they are erect, but each of these types shows variants among which may be mentioned the Indian-club form in *Trichosacme*, the oblong or sausage-shape not uncommon in *Hoya*, and the pointed elongated form in some species of *Oxypetalum*. Commonly in the Ceropeginae, and rarely elsewhere also, the pollinia have some kind of hyaline, and usually yellow, margin, which may be a slight pellucid line on the inner side as in many Ceropegieae, or apical, or basal or restricted to the upper inner edge, as in several genera, or in the form of two downwardly directed tails from the upper inner corners, as in various Stapelieae. The pollen forming the pollinia is said to have a single membrane only.

4. The translator-arms may be present or absent and vary from structures which are quite small and thin, and considerably less bulky than the pollinia, to comparatively large expansions which, in area at least, are much greater. They are commonly described as twisting after the release of the pollinia.

The arms, together with the corpuscula, are formed from a gummy secretion of the stigma-head, just as are the corresponding structures in the Periplocoideae, and this being so their least specialized condition is probably that seen in *Asclepias*, where they have the shape of a narrow more or less flattened ribbon. Sometimes they are longer and bent in the plane of the pollinia. Shorter arms are usually associated with erect pollinia and are often much broadened, and, as in the Stapelieae, may be as extensive as the pollinia and variously winged and armed. Expanded translator-arms are elsewhere best seen in the form characterizing *Oxypetalum* and its associates in which each arm consists of a large, sharply horned or toothed part and a much smaller neck, but the extreme is *Stathmostelma*.

5. The corpuscula (this seems to be the name least open to objection) are generally quite small, narrow oval or ovate in

FIG. 124. *Ibatia maritima:* plan of flower, × 5

FIG. 125. *Ibatia maritima:* vertical section of flower, × 5; pollinia × 10

outline, and much smaller than the pollinia. Sometimes, as in *Secamone,* they are cordate or even broader, or they may be long and narrow as in many of the Marsdeninae, sometimes with the translator-arms attached to the upper rather than the lower end, but generally speaking there is little variation except in the case of *Oxypetalum* and its associates. Here many species show a great development of size and also some modification of shape so that the whole has been likened to a cricket bat in outline. The best examples are those of *Calostigma,* where the "bat" is at least six times as big as the pollinium, and of *Oxypetalum strictum,* where it has a gutter form interestingly reminiscent of some of the translators in the Periplocoideae.

Stigma-head. In the Cynanchoideae the stigma-head (*narbenkopf* of Schumann, *stigmatic crown* of White and Sloane, *style table* of Corry, *stigmate* of Dop, etc.) is an integral and important part of the gynostegium to which the stamens are fused, and which, as in the Periplocoideae, secretes the translators. About it there are two points of special interest which must be considered carefully here, both involving the question of the position of the actual stigmatic surfaces.

There is no reason to doubt that the events which make up the story of pollination as usually described in the Cynanchoideae and, *mutatis mutandis*, in the Periplocoideae, take place, as indeed can not infrequently be confirmed by observation, but whether this process is in fact the necessary and invariable precursor of fertilization is held by some to be uncertain. The process described results in the pollen being deposited in the interstaminal chambers and more or less in contact with the *under* surface of the stigma-head, and the actual receptive stigmatic surfaces are usually described as being five localized patches in this position, which incidentally, is, in view of the design of the head as a whole, the only site possible. At the same time they are rarely referred to in any detail and are certainly not conspicuous, and it is difficult to avoid the impression that the statements about them are often, at any rate, more assumptions than anything else. Those who doubt their existence and function suppose that there is in Asclepiads direct self-pollination, the pollen germinating in the anthers and growing directly into and through the tissues of the stigma-head along the shortest route to the ovules, a passage little, if any, longer than that requisite if the normal pollination story is accepted. In this connection it is of interest that cleistogamous flowers have been reported in *Hoya, Stapelia* and other genera. How true these contrasting stories are will be shortly considered in the next chapter but their mention here does lead to a point which may explain some of the uncertainty about the position of the stigmatic surfaces, namely that if the pollen is placed, in masses, in precise contact with the stigma-head there may be no specialized receptive tissue, in the ordinary sense of the word, at all, and hence, in other words, no ordinary stigmas.

One curious situation arises from the peculiar construction of the stigma-head. There are only *two* carpels in the flower, separated as far as the dilation of the stigma-head and sometimes even further up, but there are *five* spots at which pollen may be deposited. One of these spots may lie on the plane between the carpels, but to any one of the remaining four (and quite possibly to all five) one carpel will be of more direct access than the other. Moreover, pollen can be deposited on only one spot at a time. If all five spots are eventually pollinated seed is likely to be set in both carpels, but if only one or two spots are pollinated, and especially if these are

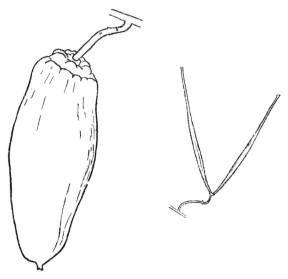

FIG. 126. Fruits of *Araujia hortorum* (left), × ½, and *Ceropegia woodii*, × ½

adjacent, it is more likely that only one carpel will set seed. This may well be the reason why it is so common to find that only one of the two carpels in a flower develops further.

These considerations make the question of the shape of the stigma-head as a whole, and especially of the upper parts of it, one of peculiar interest. Most commonly this upper part of the stigma-head is flat or nearly so, as is general in the Ceropeginae and Gonolobeae, but elsewhere in the Cynanchoideae the top of the head may be found elongated and exposed to almost any degree. In about thirty genera this elongation is such as to merit the description "conical", "long pyramidal" or even "cylindrical" or "clavate", with various qualifications, and in nearly all these it is more or less bilobed at the tip. In *Conomitra* the prolongation is said to be twisted when dry, and illustrations of *Glossostephanus* show it as a stout tapering column rather reminiscent of the whole gynecium in some other families. In some twenty genera the terms used are "rostrate" or "beaked" and these, which are also usually more or less bifid, show even more elongation. In several cases the stigma-head narrows above into shorter or longer branches which, as in *Araujia hortorum* (fig. 106), various species of *Oxypetalum*, *Mitostigma fiebrigii*, *Widgrenia* and *Morrenia grandiflora*, are usually two in number, but in *Schistogyne* there may be as many as seven. Sometimes the design of the stigma-head is even more particular, as in *Calostigma*, where there is a cupulate frill below an apical bifurcation; or in *Aphanostelma volcanensis* and *Rothrockia*, where there are lateral appendages to the prolongation. Most noteworthy perhaps are those cases where the stigma-head is flat but bears, arising abruptly from its centre, clavate or branched emergences, as is well seen in *Ibatia* (fig. 125). It is to these more particularly that the term "styliform appendages" is applicable. In many of these and of the other types the apical portion is often papillate in the manner of an ordinary stigma.

Whatever the truth about the pollination of the Asclepiads may be, it seems clear that these processes and prolongations of the stigma-head, which be it remembered occur in only a small minority of the group, are not concerned with it, because whether there is effective insect pollination in the way described above, or whether there is direct self-pollination, only the under sides or lower lateral parts of the stigma-head can be involved. The great interest in the variation of the upper surface of the stigma-head is, therefore, that with increasing elongation and bifurcation it comes to have, more and more, the superficial appearance of the normal state of affairs in less peculiar families, that is to say it becomes more and more like a true stigma. Are we to suppose that the true stigmatic surfaces have migrated from the very tops of the styles, leaving these functionless, in which case one might expect the evolutionary tendency to be towards their repression, or are we to suppose that these elongations have developed secondarily from the flat surface of the stigma-head, thus forming "false" structures, namely structures which appear to be stigmas but which are not? Whichever the answer the implications of it are of great interest, and become even more so when there are recalled the "false stamens" in *Barjonia*, and the peculiar arrangement and appearance of the coronal lobes in some Stapelieae.

Fruit. The interest of the fruit in the Cynanchoideae is not so much its variation, though this is not inconsiderable, but in the circumstance that it is of a design peculiar to the Asclepiads and some of the Apocynads.

The gynecium of the flower consists of two carpels, and although their styles are united above during anthesis by the stigma-head their ventral parts are distinct and normally ripen into pairs of separate large, green, spindle-shaped, soft-walled follicles which dehisce lengthwise and which are without close parallel outside the Apocynales. Each carpel typically contains many ovules, and each follicle large numbers of flattened, beautifully hair-tufted seeds, very regularly and exactly packed. The dehiscence is ventral, along the line of the axile placentation and as it proceeds the hairs of the seeds expand and the whole seed-mass, or a considerable part of it, held together by the interlocking of the tips of the hairs, emerges into the air and is gradually blown to pieces.

One strange feature is that very often only one of the two carpels ripens, so often and consistently indeed, that it seems in many cases to be almost a specific character. Allied to this is the fact that the total number of follicles ripened in any one multi-florous inflorescence is often extremely limited, and may even be no more than one. On the other hand some species, in conditions of cultivation at least, habitually ripen much larger numbers and produce great quantities of seed.

The general range of characters in the follicles is what might be expected, being based on such variants as widening and narrowing, bluntness and sharpness, straightness and curvature, and so on, while they may also vary from parallel to divergent, from smooth to warty or spiny, from glabrous to hairy, and may be leathery, woody, papery, or fleshy in consistency when fully ripe. Extreme states are neither numerous nor very conspicuous and may be sufficiently illustrated by the mention of the following examples, namely, large fleshy, velvety fruits nearly a foot long in *Marsdenia mollissima*, and not much smaller in *Morrenia odorata*; long cylindrical bean-like fruits in *Ceropegia candelabrum*; fruits covered with pubescent warts and appendages in *Prosthecidiscus* and *Ibatia*; fruits covered with many deep, broken, sinuous wings in *Dregea abyssinica*; fruits with circumferential overlapping scales or wings in *Philibertia schreiteri*. There is only one striking anomaly in the whole group, the genus *Emicocarpus*, in which the fruits are indehiscent and three pointed, like those of *Emex* in Polygonaceae, and have only one or two seeds. This genus is unusual in other ways also, e.g. leaf and corona, and is thus an interesting counterpart, in the Cynanchoideae, of the aberrant genus *Finlaysonia* among the Periplocoideae. No doubt there would be more to say here about fruits were more information available, but in a surprisingly large number of the rarer genera and species they are still undescribed.

From the evolutionary point of view it is interesting to note that although there is such a consistency of basic design in the fruits of the Cynanchoideae the scale of variation just described is sufficient to produce results of remarkably different appearance, as can be seen in fig. 126.

Seeds. The chief seed variations are in the number in each carpel and in the presence or absence of hair-tufts. *Stigmatorhynchus* is said to have only one, hairy, seed in each follicle; *Emicocarpus*, and probably *Eustegia*, have only one or two hairless seeds; *Madarosperma* has a few, without hairs; and in *Sarcolobus*, or some species of it at least which are plants of coastal mud, there are hairless seeds, so that these plants afford a special parallel with *Finlaysonia*. In *Dorystephania* and *Ischnostemma* the seeds have narrow wings, and in *Pseudibatia* they have broad, dentate borders. It may be noted, though it is scarcely of direct concern here, that the ovules of the Cynanchoideae have been described (by Dop and others) as having an unusual and simplified organization.

NOTE ON LITERATURE

The main sources of information about the Asclepiads in general are five:
1. The monograph by K. Schumann in the first edition of the *Pflanzenfamilien*, 1895, which incorporates earlier work and which still remains the only fully integrated well-illustrated revision of the family as a whole.
2. A. White and B. L. Sloane's fine illustrated monograph of *The Stapelieae*, second edition, Pasadena, 1937. This deals with what, to

many, is the most interesting part of the family and contains also some more general information.

3. A small volume by T. M. MacFarlane entitled *The Evolution and Distribution of Flowering Plants. Vol. I. Apocynaceae; Asclepiadaceae*, and published in Philadelphia, 1933, which gives a general cursive account of both families. It traces hypothetical phylogenetic relationships in great detail and contains much useful information, but is so speculative that its wider value is lessened.

4. The generic descriptions, and especially the complete analytical generic key (vol. 8b) in A. Lemée's great *Dictionnaire déscriptif et synonymique des genres de plantes phanérogames*, 1929–1943–

5. Certain of the larger illustrated floras, especially the now elderly *Flora Brasiliensis* by C. F. P. Martius and the sumptuous new *Flora Argentina* by H. R. Descole.

Among shorter publications the following, particularly, are relevant here:

1. Corry T. H.: "On the mode of development of the pollinia in *Asclepias cornuti*, Decaisne." *Trans. Linn. Soc.*, Ser. 2, **2**, 1881–87.

2. Corry, T. H.: "On the mode of development of the gynostegium in *Asclepias cornuti*, Decaisne." *Trans. Linn. Soc.*, Ser. 2, **2**, 1881–87.

3. Delpino, F.: *Sugli apparecchi delle fecondazione nelle piante auto-carpae (Fanerogame)*. Firenze, 1867.

4. Demeter, K.: "Vergleichende Asclepiadaceenstudien." *Flora*, **115** (NS 15), 1922.

5. Dop, P.: "Recherches sur la structure et le développement de la fleur des Asclépiadacées." *Comp. Rend. Acad. Sci.*, **136**, 1903; also published as *Théses presentées a la faculté des Science de Paris*, Sér. A., No. 447, 1903.

6. Falconer, H.: "On a reformed character of the genus *Cryptolepis*, Brown." *Trans. Linn. Soc.*, **19**, 1845.

7. Good, R.: "An Atlas of the Asclepiadaceae." *New Phytologist*, **51**, 1952.

8. Groom, P.: "On *Dischidia rafflesiana* (Wall.)" *Annals of Botany*, **7**, 1893.

9. Hildebrand, F.: "F. Delpino's observations on the pollination mechanisms in the Phanerogams." *Bot. Zeitung*, **25**, 1867.

10. Lückhoff, C. A.: *The Stapelieae of Southern Africa*. Cape Town and Amsterdam, 1952.

11. Pearson, H. H. W.: "On some species of *Dischidia* with double pitchers." *Journ. Linn. Soc. (Bot.)*, **35**, 1901–1904.

12. Scott, D. H. and Sargant, E.: "On the pitchers of *Dischidia rafflesiana* (Wall.)" *Annals of Botany*, **7**, 1893.

13. Woodson, R. E., jr.: "Studies in the Apocynaceae: a critical study of the Apocynoideae." *Ann. Missouri Bot. Gard.*, **17**, 1930.

14. Wydler, H.: "Morph. Mitteilungen. 1. Infloreszenz in *Vincetoxicum* . . ." *Flora*, **40**, 1857.

CHAPTER EIGHT

THE ASCLEPIADACEAE—II

THE account of the Asclepiads in the last chapter was designed to fulfil two purposes, a more general one of illustrating the content and composition of one of the larger constituent groups of the Flowering Plants, and thus the kind of evolutionary problems that any one of these is likely to present, and a more particular purpose of calling attention to those special features of the Asclepiads which show some, at any rate, of these problems in a peculiarly inescapable way. Much condensed though it was the account of these plants was lengthy, and it is better to consider the theoretical aspects of some of the facts while they are still fresh in the readers' minds. In total these problems are so many that no brief treatment can hope to do justice to all of them, and attention, in the following pages, is therefore concentrated, first on what must, on any reckoning, be held to be the problem-in-chief, and then on three selected other problems which require less lengthy consideration.

Problems of the Pollination Mechanisms

The outstanding evolutionary problem in the Asclepiads is to explain the origin and development of their peculiar pollination mechanisms, which are uniquely different in each of the two sub-groups. In the Periplocoideae, as the reader will remember, the pollen, which is in discrete or agglutinated tetrads, falls on to or into special structures, the translators, which are formed from a secretion of the cells of parts of the stigma-head and which are so designed and situated that they can be removed from the flower by adhesion to some visiting insect. In the Cynanchoideae the pollen is aggregated into definite pollinia which are joined together in pairs by yoke-like translators which are secretions of the cells of parts of the stigma-head and which are so formed and so attached to the pollinia that when they are removed from the flower by insects these drag their pollinia with them. As far as the writer is aware there is neither statement nor suggestion that transfer of pollen *inter flores* is, in these groups, ever effected in any other way, and this being so there is at least negative justification for assuming that they have no other means of either geitonogamy or cross-pollination, and that the normal methods by which these are effected have either never been possessed or have been lost.

There can be little reason to doubt that, while the Asclepiads are a homogeneous group as a whole in a phyletic sense, being all derivatives of one and the same stock, each of the two subfamilies is even more so. Furthermore there is no reason to suppose that the asclepiad type of floral organization represents any very early and generalized angiosperm condition, or that the more ordinary types of today are derived from it, and it would therefore seem that this peculiar condition has become, at some point in time past, substituted for something more stereotyped. Hence any enquiry into the subject must concern itself primarily with *how* and *why* this may have happened.

It is a particular aim of this book to avoid theories or assumptions concerning phylogeny and the relative primitiveness of types, but there are some subjects, of which this is one, which cannot usefully be pursued without the help of at least a working hypothesis of this sort, though even here it must be the simplest possible. It can scarcely be denied with reason that all the facts described in the last chapter, especially in its earlier pages, suggest that if the Asclepiads and Apocynads have evolved, they have come from or belong to a common stock, and that those of the latter which now exist more closely resemble this earlier stock than do the former. If this view be accepted then it is possible to gain a valuable and important impression of the nature of the changes which may have produced the specializations of the Asclepiads.

As we have seen, the difference between the two families today, though it would seem to be functionally complete, is morphologically even smaller than is generally realized. Such features as the presence of small glands on the inside of the calyx; the presence of a corona; the occurrence of tetrad pollen; and even the secretion of viscid liquid from the stigma-head, all occur in the apocynads and when seen there singly or in only partial combination excite little comment. When, however, as happens on occasion, these occur in greater or almost complete combination, they result in a floral condition that differs very little from that of some of the Periplocoideae. Indeed, it is only in the presence of *functional* translators that this group is really to be distinguished, and since it has been claimed that non-functional homologues of these occur in *Apocynum*, even this last gulf between the two may be only the difference between incipience and full development in a single character.

Between the Apocynads and the Cynanchoideae there is a much wider gap because of the occurrence in the latter of the definitive pollinia and highly specialized translators, of which there seems to be no trace in the Apocynads, while for the same reason there seems to be no bridge across the gap between the Periplocoideae and the Cynanchoideae, though it is important to remember in this connection that many of the genera in the latter are so little known that much about them may remain to be learnt. However, Demeter has maintained that this more complex translator of the Cynanchoideae can be derived, mainly by a process of folding, from those of the simpler forms of the Periplocoideae. It would seem therefore that the three groups Apocynaceae, Periplocoideae and Cynanchoideae, are expressions of the same general line of descent; that the morphological discontinuities between them are least between the Apocynaceae and Periplocoideae; somewhat greater between Periplocoideae and Cynanchoideae and most marked between the Apocynaceae and the Cynanchoideae. Because of this more attention than is warranted by the mere size of the group will be paid here to the Periplocoideae, and it will be important to remember throughout that what may be said of that group is even more widely true of the Cynanchoideae.

The close link between the Apocynads and the Periplocoideae simplifies our enquiry in one way because, if it had not existed, it would have been necessary to try to decide whether the Periplocaceous condition could or could not have arisen suddenly from an ancestral form with what may be called the normal kind of flower. It must be remembered that there is no logical reason why, indeed, this may not have happened even granted the present closeness of the two groups, and it may be more than once, but in view of what has been said, there is greater justification for assuming the simpler explanation that the gulf has been bridged at its narrowest point, and whatever difficulties may be involved in this opinion, they are at least smaller than those which would arise from any other supposition.

To reduce the problem to its simplest expression in this way is really to accentuate it, for the residual gap remaining between the two after all has been said, is the real crux of the matter, because it involves a fundamental alteration of biological mechanism. How, may it be supposed, did this change of function come about, and how, in particular, did the secretion of viscous matter by the tissues of the stigma-head become co-ordinated into the production of functional structures for the removal of pollen?

In trying to arrive at some answer to this question it is salutary to remember what an astonishing chain of circumstances is needed before the functional condition of the Periplocoideae, and still more of the Cynanchoideae, is reached. The stigma-head, or its precursor, must possess the power of secretion; this secretion must be of such a sort that it hardens in, and is resistant to, air; the distribution of the secreting cells must be such as to give the secretion an appropriate shape; the secretion must have a particular spatial relationship with the stamens; the surface of the translators must be such as to permit the retention of the pollen; the translators must become completely independent of the tissues that secreted them; and, most astonishing of all, the translators must become differentiated into distinctive parts, none of which has any particular value in the absence of the others. Add to this that in the Cynanchoideae the pollinia must become firmly attached to the translator-arms and that the anthers must antecedently dehisce to allow this.

Among these and many other components in the passage from the apocynad state to that of the Asclepiads (assuming such passage to have taken place), two stages must have been more crucial than the rest, one earlier and one later. The first was the modification by which the stigmatic or stylar tissues came to secrete, not merely the sugary solution common in such tissues, but a more tenacious substance of structural potentiality. This was presumably the consequence of a definite chemical change expressed ultimately in the production of some new, or at least different, kind or kinds of molecules. The second stage was the final achievement of functionalism in the translators. At one stage they must have been incapable of acting as pollinating agents; at the next they must have been so capable. Three conditions in particular were attained—the translators became easily detachable; they became capable of retaining the pollen; and they became differentiated into parts. The third of these especially is a critical change which it is hard to visualize.

All these changes must have been controlled either by external factors, including factors of the environment in its widest sense, or by internal factors, or by some combination of the two. In so far as chemical change is inevitably some expression of metabolic processes there is no great difficulty in imagining one way in which external factors may have played a part, but in what manner these may have made themselves felt in some of the other sorts of changes, is difficult to see. What can have been the relation between a metabolic change and the initiation of a structural alteration, and how can the one have come to expression in the other? It may be argued that given chemical change brought about by the operation of external factors, almost anything may happen, but this does not help us to see the actual process. On the other hand, if internal factors are considered to have been of greater importance, then it clearly becomes desirable to specify as nearly as possible what these may have been, and in fact it is hard to do more than infer the operation of some innate or inherent tendency to change by which a prior condition becomes converted into a subsequent state, and this again is not very illuminating. Indeed, this precise problem of the manner in which metabolic change (which may be due to a variety of factors) becomes converted into structural reorganization is near the heart of the whole problem of organic evolution, and this case of the Periplocoideae presents a most remarkable exemplification of it. Could we but reach an understanding of this particular problem we should have progressed far in our understanding of organic evolution.

There is in this problem something of a true dilemma which concerns the fundamental issue of how far, in evolution, the conditions at any one given time may determine those of a later date, for, if it be supposed, on the one hand, that one state may be the inevitable precursor of another, then evolution would appear to be, in some measure at least, determinate, while, on the other hand, if it is supposed that all evolutionary change is inconsequential, then resemblance may be purely fortuitous and without real meaning. The significance of this dilemma is well seen in the Asclepiads. For working purposes the assumption was made above that the essential early step which paved the way, and without which many later stages would have been impossible, was the alteration in the nature of the secretion of the stigma-head, but acceptance of this assumption at once raises the question of the relationship between this and later developments. Was this primary change the first of a predetermined series destined inevitably to culminate in the hitherto unexpressed condition of functional translators, or did these translators arise merely through a succession of unrelated chance changes subsequent to the primary change, but having no direct causal connection with it? This is the kind of puzzle which recurs again and again in considering the nature of evolution in the Flowering Plants and to it there is apparently as yet no satisfactory answer. It ought, however, to be noted that neither of the two suggestions made is simpler than the other in its implications, and the establishment of the truth of either would have the widest repercussions on biological thought.

There is a rather similar dilemma in respect of the possible effects on evolution of environmental factors. To deny entirely that factors of this kind have influenced the evolution of the Periplocoideae would be to cast away the potential aid of one important thesis in most evolutionary thought, but it is nevertheless extremely difficult to see how such factors can have operated in producing those particular characteristics which distinguish this group of plants. This is not to suggest that such factors have not operated, but simply to stress the implications of believing this to be the case. It is true that the conception of environment may be made very wide, and that some conditions of this sort may, by their influence, cause some of the changes which are part of evolution can scarcely be gainsaid, but to suggest exactly how environmental factors may have exerted influences resulting in and causing the unique floral organization of the Periplocoideae, and *a fortiori*, that of the Cynanchoideae, is quite another matter.

But whatever the difficulties may be in the problem of *how* the evolution of the Asclepiads may have come about, those involving the question of *why* it occurred are even more profound. Whether, indeed, in the present state of our knowledge anything like a useful answer to problems of this kind is possible may be doubted, but at least it is valuable to comment upon them and to try to indicate the relative merits of differing opinions. The core of the problem is the question of the biological "efficiency" or "value" of the asclepiad floral mechanisms as compared with others of more usual kinds, and this difficult problem involves especially two matters, the infrequency of the Asclepiads in general, and the efficacy of their pollination.

In studying the Asclepiads one is constantly faced with minor difficulties that arise from the lack of precise information about many of the members of the group, and in quite a number of cases it would appear that species and even genera have been described from very few collectings and even from only one. A striking case of this is the very distinctive genus *Frerea*, the only Stapeliad possessing persistent leaves. This was found and described from the specimens collected on a single hillside near Poona in 1864. This can hardly be considered one of the more remote parts of India and yet the plant was, if accounts are correct, never seen again until about 1930 when it was collected once more in the same place. Now it is true that the adequacy of collected specimens to some extent reflects the adequacies of collectors as well as their distribution, but making due allowance for this it seems clear that our limited knowledge of many Asclepiads derives partly at least from the fact that the part these plants play in the general scheme of things is less conspicuous than that of many other families and that many of them are indeed *rare*, using that term to include not only scarcity of individuals but also extreme restriction of range. Some of the temperate herbaceous forms, especially among those in North America, are somewhat gregarious and may occur in considerable quantities, but the tropical and sub-tropical forms, which comprise among them by far the greater variety of characters, appear, with comparatively few exceptions to be uncommon plants. Without a first-hand acquaintance with more tropical and equatorial asclepiads so wide as to be scarcely attainable under the most favourable conditions, it is hard to say how real this general impression is, but a number of indications confirm it, among them the paucity of published records; the scarcity of specimens in herbaria; and the number of incomplete or dubious statements, while it is confirmed by some remarkable positive statements. These can best be illustrated by reference to that section which is best documented, namely the Stapelieae, one example of which has just been mentioned, but it is important to remember that by implication the same is equally true of many other parts of the family.

One of the most striking instances here is that of some 36 species of South African Stapelieae originally collected by Masson in the late eighteenth century. Of these six, and perhaps three others also, have not (*vide* White and Sloane, 1937) been seen again and are known only from his notes. That these were perfectly genuine appears likely from the fact that nine others of his species *have* been rediscovered, but only within comparatively recent years. It is to be noted that other species of his description are common species which have been known throughout, and also that his collectings were made in areas which have long been easily accessible.

Another very remarkable circumstance is that of all the species listed by White and Sloane, over 40 per cent., or upwards of half, are recorded only from single localities, and are presumably known only from these. Nor is it that these are from remote parts and quite a number of them are from the more familiar parts of the Cape Province. Not only this but some of these species are known only from single plants.

It may well be argued that there are other tropical families in which an equal proportion of members are similarly rare and little-known, but it is not the purpose to maintain here that the Asclepiads are outstanding above all other groups in this respect. The facts which have been cited are rather intended to support the view, to which we shall revert later, that the Asclepiads can in no real sense be called either abundant or aggressive in comparison with many other families.

The situation with regard to pollination is difficult to summarize briefly because there are so many items of interest in it that one is tempted to expatiate upon it beyond the space available. Let it suffice then to make three statements. First, there is no reason to suppose that pollination *inter flores* ever takes place except through the agency of insects and by the methods described on pp. 210–215 above, and the actual carriage of pollinia in this manner can easily be demonstrated by catching insects in flight with the pollinial apparatus attached to them. Second, the suggestions of direct positional self-pollination which have been made from time to time appear to have no general application though the possibility of this in any particular circumstances cannot be entirely excluded. The point to remember here is that if self-pollination should be demonstrated to occur widely it would become, not easier, but more difficult to account for the complex translator mechanisms. Third, though often enough in nature very few ripe fruits are produced, the present writer can, in company with others, confirm as a result of personal observation that sometimes enormous quantities of fruit containing viable seeds are quite normally produced in circumstances in which the possibility of self-pollination seems excluded. In short there seems no valid reason to doubt that insect pollination of the characteristic sort described is normal in the Asclepiads, and adequate for the biological purposes of these plants.

Being so closely associated with pollination, the question of hybridism in the Asclepiads is one of great interest. References to hybrids among the Stapelieae are frequent but elsewhere in the family they are almost unknown, and this last fact, coupled with the general necessity in asclepiad pollination that "the insect must fit the flower" much more nicely than in most ordinary families, inevitably raises doubts as to the truth of the matter in the Stapelieae. It is always difficult to establish the fact that a species or individual is a hybrid but it is nevertheless noteworthy that there is scarcely to be found one categorical statement that any member of the Stapelieae is of this nature, they are simply assumed to be so on the grounds that they are intermediate in characteristics between known species. Other enquiries, too, have failed to reveal any substantiation of these claims. Again some Stapeliads are extraordinarily polymorphic in form and corolla colour, notable among these being *Stapelia variegata*, a common and often cultivated species, and all in all there seems good grounds for doubting whether the Stapelieae concerned are in fact hybrids. The importance of the point is that if they are not hybrids then they must be ordinary species or varieties, and if this is so then the facts which have led to the belief that they are hybrids point strongly to the explanation that they are forms (and it may be imagined in most cases new forms) in which there have appeared fresh combinations of characters already present and expressed individually or in other combinations in the lineage of the group. Taking this explanation further towards its logical conclusion it would seem distinctly possible that these reputed hybrids are in fact new forms, and that this indicates that the group of the Stapelieae is in active evolutionary phase by mutative recombinations of characters. It may well be also that the extreme rarity of some species indicates that they are evolutionarily new species.

But this has taken us rather far from our main theme, the question of the "value" of the Asclepiad mechanism, and to this we must now return, fortified however by the knowledge that the members of the family cannot be considered notably aggressive or abundant; that they are often extremely restricted in range; that the pollination is effective but that in nature it does not very often produce copious fruits; and that hybridism is very unusual except possibly in one section of the family. To these facts may be added the observation that the Asclepiads do not, with rare exceptions, show anything in the way of exceptionally wide tolerance to edaphic or climatic conditions; that they have little power of vegetative propagation; and that they provide practically no weeds of the more objectionable kind, although one or two of the herbaceous forms are widely adventive. It is also to be remembered that they are predominantly a group of climbing plants. All in all then they give the impression of moderation on nearly all counts and especially on the ecological side and it is therefore not at all easy to maintain that they are in the more usual sense of the phrase notably "successful", or that they possess absolute advantage over other types.

Here, of course, there is the real difficulty of defining, or at any rate suggesting, what may constitute "success" or "advantage" in organisms of this kind, and on this matter there are some important comments to be made. Whether indeed the human mind is capable of making a true estimate of this kind may be doubted, but in such case some current opinions need even more careful examination. It would seem to be a reasonable major premiss, that the continuity of the race, in the sense of the continuance of viability, is to be regarded as the prime biological aim, but what the ideal form of this may be from the overall point of view, is by no means certain. Because of the lack of any other ready yard-stick, and also as a reflection of human experiences, there is a natural tendency to express opinions in terms of quantitive values, but this is not altogether satisfactory. Thus it is commonly implied, even if it is not explicitly stated, that species which maintain themselves in great quantity over wide areas are more successful than others. Certainly it cannot be denied that such plants are extremely successful in doing the particular thing of maintaining themselves thus, but whether this is what constitutes real biological success may be questioned.

It may, indeed, be wrong and misleading to attempt to measure biological success in this quantitative way at all. It may be, that in the widest sense, there is virtue in survival on the smaller rather than on the larger scale, for this would at least effect economy of material and would, which is perhaps more important, reduce the problems of spatial competition which, in plants at least, constitute so much of what is rather picturesquely called "the struggle for existence". It may even be that if, as is believed, some form of evolution is the fundamental biological process, the real successes are those species which have the briefest existence in time and which most rapidly change, by the processes of evolution, into something else. Rapidity of evolutionary effacement may be a surer mark of biological success than any long immutability, and mutational fertility of a stock almost certainly better than any degree of reproductional fertility among the individuals. Perhaps it is even not too fanciful to regard mutability as being the true basic function of any aggregate of organisms. Should this indeed be so, then continued existence on the grand scale to the possible exclusion of other types would seem to be no advantage unless the stock concerned has a more than average power of mutation, and the continued maintenance of enormous numbers of virtually similar individuals in any species may be evidence of the lack of this power.

These difficulties do not perhaps loom so large if the problem is posed in terms of advantage rather than of success, but this phraseology has its own puzzles. If, for example, it is supposed that there exist certain types of structure and behaviour which confer advantages on the organisms possessing them by furthering their racial continuance, then the operation of any kind of natural selection would seem to require a convergence of evolutionary trends towards these conditions, which must, *ex hypothesi*, be comparatively limited in number. Not only so but it would seem, on the same grounds, unlikely that parity of advantage would result, but rather that a limited number of conditions would be more advantageous than any others, or even, to press the argument to its logical conclusion, that one only would, in this respect, transcend all the rest. Were this the case it might be expected that in any great group of organisms such as the Flowering Plants there would be an overall impression of unification and of convergence to one particular state. But the facts do not reveal this, and it is differentiation and the multiplication of novelties, rather than unification and the suppression of originality, which appears as the underlying theme of nature. The overwhelming impression is that a multitude of different, quite unconvergent, conditions may confer on their respective possessors the necessary amount of biological self-sufficiency, and that the multiplicity of the Flowering Plants for example is to be interpreted, not as a vast competition for supremacy, but rather as an expression of the great number of ways in which this sufficiency may be obtained. Nor should this cause surprise, for evolution, since it is in essence change, can scarcely continue on a centripetal or contracting plan.

The Asclepiads are obviously of special interest in relation to this argument because they provide, in the groups Periplocoideae and Cynanchoideae, two outstanding instances of what appears to be centrifugal departure in evolution and of the production of novel conditions, evolution in the stock of their predecessors having led to two conditions of organization and mechanism which are not quite paralleled elsewhere in the plant kingdom and which depart in notable degree from the generality of form.

But it may be argued that the conception of one or a few ideal forms in nature is too narrow and that very many different states are to be regarded as equally desirable or satisfactory, but this serves only to complicate the issue by modifying the conception of selection, and also introduces philosophical problems of choice and of the estimation of real value which can only be approached with difficulty. Rather it seems necessary to assume that imperfection, in the sense of something short of the absolute best, is the normal condition in nature. Certainly observation goes to show that success, if it exists at all, is after all only of moderate degree. Never does one type attain such superiority as completely or even largely to swamp all others on a wide scale. Local dominance may occur in many detailed circumstances, especially where man takes a hand, though even here real gregariousness is not very common, and there is of course the secular replacement of dominant groups in the course of geological time, but neither of these is an example of the process under discussion.

It may also be maintained that the theory of natural selection particularly permits the view that variations which are neither useful nor injurious need not be affected by selection, and that such a character as the development of translators in the Asclepiads is a variation of this sort, but even if this is so it merely removes one conceivable explanation of the facts without substituting any other for it. It is certainly hard to see in what way the remarkable changes which have occurred in order to produce the members of this family have conferred upon these plants any kind of special superiority or advantage over other plants, and this being so, to see what leading part natural selection can have played in their production.

The Asclepiads are of particular interest also in regard to another troublesome aspect of selection, the difficulty of assigning any selective potential to characters in an incipient or pre-functional state. There are really two aspects of this problem. The first is that if the suggested sequence of events by which the translator-mechanism came into being is correct, then this form of pollen transference must have been derived from one of the more ordinary kinds, and it has already been shown how many constituent changes must have been involved (if gradual evolution is supposed) before the final functional state of the new organization is reached. How, then, can the change have been effected? Clearly in the course of the process the old method became superseded by the new, but were the two *régimes* consecutive and continuous in time, or did they overlap, or was there a gap or interregnum between them? If the first, how could the change from a long established method to a new and untried one depending on structures which until that moment had had no function, have come about? If the second, then at some stage the plants concerned must have possessed both methods of pollen transference at the same time, and why, in such circumstances, there should have come about a change of balance from one to the other is not easy to understand. If the last, then biological continuity across the gap could presumably have been maintained only by some form of direct self-pollination, and the loss of the original function would be even more inexplicable.

The second aspect brings us back to the problem of the possible way in which a stigmatic or stylar secretion might become a functional translator. Earlier concern was chiefly with the possible steps in such a process; here it is the more underlying causes which are at issue. Presumably in the course of time, and by steps which cannot now be recalled, something that was merely a watery secretion became organized (to use that word in one of its more exact meanings) and, ceasing to be no more than a subsidiary product, became an integral part of the floral mechanism. Selection could hardly have influenced a non-viable secretion directly; it could only affect it by influencing the living plant itself in such a way that it responded by some metabolic change which resulted in a new kind of secretion. But even so, how could any such change in a non-viable structure be orientated towards the fulfilment of a condition whose ultimate function will only make itself apparent at some future time? In short, how can selection act upon a character which has no function, except possibly to eliminate it, and how can such a character become functional in the absence of selection pressure? These are old enigmas which formed the bases of some of the earliest criticisms of the theory of natural selection, and they have become pushed aside in the course of time, but they are nevertheless real, and a study of the Asclepiads gives them a fresh and urgent currency.

But there is still one more profound question that awaits some answer. How far is it justifiable to incorporate the conception of anticipation into evolutionary theory? Are the successive states in an evolutionary process simply a series of unrelated changes or is there some thread of continuity of purpose running through them from end to end? Is it to be believed that through the production of a particular kind of secretion by the stigma of an ancestral type, the ultimate establishment of the Asclepiads became an inevitable and foregone conclusion? These are difficult questions indeed but these plants make it impossible for us to ignore them.

To recapitulate, it would seem that any proposition that the peculiarities of the asclepiad flower have been derived by the operation of some kind of natural selection, involves two premises, one that these peculiarities are advantageous, and the other that the antecedent conditions were capable of being selected. The discussion has shown how great are the difficulties in the way of accepting either of these. This being so what other explanation is feasible?

Assuming that the Asclepiads have indeed had an evolutionary origin, that is to say that they have arisen from some pre-existing type or types by a process of change, this change can be conceived in terms either of continuity or of discontinuity. The former idea is that behind the theory of natural selection and has been discussed, the latter is that behind the theory of mutation. At first the idea that the Asclepiads originated through a large-scale mutation from some other group, presumably one which was like the present Apocynads in many respects, seems to resolve many difficulties, but we must be cautious and review its implications with some care. For one thing nothing but a single large mutational change bridging completely the gulf between the pre-asclepiad type and the fully functional asclepiad type will serve the purpose, because anything less than this would require to be repeated by one or more successive changes, and the distinction between such a series and the events which are commonly comprehended in the idea of natural selection would be very difficult to draw, and would scarcely serve to solve our problem. From the mutational point of view therefore the question really resolves itself into whether or not it is reasonable to imagine that the peculiarities of the Asclepiads were attained, in the course of a single major mutational change, from a stock which did not possess them. To debate the powers of the human imagination would probably be an unprofitable and certainly an unwelcome, task, and it must be held sufficient here to point out that in two directions at any rate it would certainly strain this imagination to accept a mutational explanation of the facts. One of these is the difficulty of understanding, in the absence of any precise knowledge about the antecedent condition, how such a sudden change of this kind and magnitude could actually shape itself, and how the changeover to the new condition could actually be effected; the other is to see in the mind's eye how, even if this change were brought about, the new condition could at once find itself in proper and functional harmony with the rest of nature. Thus, any explanation based on the theory of mutation appears to be no more free from difficulty than any other, and there for the present the matter must be left.

Summing up, it may be said that the real evolutionary problem of the Asclepiads seems to be this. Here is a large group of plants, amounting in all to more than one per cent of flowering plants, which is peculiar in having pollination by means of structures called translators which, if recorded statement is correct, are formed from secretions of the stigma-head. The evolutionary development of these translators must date from certain initiations and changes in this secretion, and there are three ways in which it may be possible to explain subsequent events. One is that there were sudden and complete mutational changes from some precursor to each of the two floral conditions represented by the Periplocoideae and the Cynanchoideae. A second is that the secretion was, from the beginning, something destined eventually to become functional in two quite different forms by a process of gradual change. The third is that function was eventually achieved, by the same procedure, at some later date when the secretion had already developed beyond some critical level of organization. The first of these would if it could be demonstrated, throw much new light on the possibilities and processes of evolution, but as a theory it has obvious difficulties. The second involves the recognition of the conception of anticipation in evolution. The third requires some particular explanation of how the earlier changes, before the critical level was reached, could have come about.

The Problem of the Corolla and of the Corona

Another problem of the Asclepiads in relation to evolution is that of the elaboration of the corolla and, much more extraordinary, of the corona. One is accustomed to some degree of corolla variation in many groups of the Flowering Plants, and indeed in larger families it is the general rule, so that it is as a specially good example of something that occurs widely that it is discussed here. At the same time the most striking instances of it elsewhere are chiefly associated with the highly heterogeneous families such as the Scrophulariaceae, in which there is no underlying constancy of floral design, and to find it occurring in conjunction with such particular floral mechanisms as those of the Asclepiads is certainly noteworthy. It is notable also that this elaboration, although it is remarkable enough, is much less than it might be, for it incorporates no one of the features which most readily give variety to floral organization, namely zygomorphy, polymery and apetaly. How far these limitations are associated with, and the result of, the special floral mechanisms is itself a matter of considerable interest which must not be overlooked.

In short, there are certain quite common floral characteristics, notably zygomorphy, variation in number of parts and apetaly, which do not appear in the Asclepiads, but other and less usual features are elaborated to a quite unusual degree, almost, as it were, as if in compensation, or in substitution. Corolla ornamentation and pattern, corolla texture, and even perhaps more than either of these, corolla hairiness, manifest themselves in quite outstanding fashion. Similarly in the characters which have a more limited gamut certain relatively unfamiliar conditions are emphasized. Thus, flower colour is not particularly conspicuous in the group, though most hues occur here and there, but combinations of some of the less usual colours are distinctly prominent, while often there is a marked development of hairiness which serves to heighten what might otherwise be a sombre effect. It is notable also that, within the bounds of actinomorphy, form has been widely exploited, and is accompanied by some rather special features such as differentiation between lobes and tube, adhesion of the tips of the lobes, and curvature of the tube. To recapitulate, the asclepiad corolla is fundamentally unexceptional and restricted in form, but this is more than balanced by the development of special shapes, colours, textures and hair-coverings each of these reaching, in one type or another, extreme expression.

The problem of the corona in Asclepiads, though clearly related to that of the corolla, is basically different because a corona is a rarity among the Flowering Plants and therefore apparently a less essential structure than many other parts of the flower. It may also be recalled here that coronas among flowering plants are not all morphologically similar, though most of them are presumably elaborations of the corolla, and that, *ipso facto*, the corona is likely to show less potentiality of development than the corolla itself, which makes its extraordinary expression in the Asclepiads all the more remarkable.

To attempt to discover or define the function of any plant structure is to tread on dangerous ground but it seems a fair enough statement that the asclepiad corona has no very definite function, a view supported by the fact that it is sometimes little in evidence and may be entirely absent. It is true that many of the coronas are in one way or another nectarial, but we know from other flowering plants that floral nectaries may be of many other kinds. It is also true that in many asclepiad flowers the presence of the corona lobes does facilitate contact between the feet of visiting insects and the anther clefts, but there are many cases where this is not so, and the whole question of the relation between the corona and the pollination mechanism is one of great interest and some obscurity.

The pollination mechanism is fully developed throughout the group but the development of the corona is much less complete, and it is well to remember that, while coronal form is marvellously varied, it obviously expresses but some of the possibilities. It is, for instance, hardly ever petaloid in the way in which it might produce a condition homologous with that in the daffodils.

But this is a particular aspect of the corona and its primary interest lies in the illustration it affords of what may be called gratuitous radiate or elaborate evolution, namely evolution which appears to be unrelated to the better performance of any function and which may therefore be regarded as change for its own sake, or change which is due to some cause other than factors of immediate circumstance. The whole impression given by the asclepiad corona is of an organ in a high state of instability, with little crystallized form, and at least capable of taking, if not actually liable to take, almost any form within certain wide limits. Within this broad conception there are, it has been made plain, many more particular interests and problems but it would take us too far into the *minutiae* of the subject to pursue these here, and we must confine ourselves to stressing the wider issue. What is the basis for these extraordinary manifestations of detailed form; what is their relation to, and within, the whole broad process of organic evolution, and by what machinery do they express themselves? These are the real questions at issue with regard to the corona of the Asclepiads, and, let it be clear, in the many other similar, though perhaps less impressive phenomena in other families of the Flowering Plants.

The Problem of False Structures

Another major evolutionary problem of the Asclepiads is that of what were described in the last chapter as "false structures". Particular examples of them were noted in the corolla-like coronas of *Margaretta* and in the anther-like glands on the corolla of *Barjonia cymosa*, but there are two other more important expressions of them.

One result of the peculiar pollination organization of the Asclepiads is that their flowers are without the emergent stamens and styles which so commonly provide a radiating central feature in more ordinary families, and it is therefore of great interest that this deficiency is not infrequently made good, at least in some measure, by the detailed form of the corona. In any case, of course, simple coronas, being composed of five lobes arranged round a comparatively small centre, tend to this effect, especially when, as in some species of *Hoya* (fig. 110), they are pointed in outline and darker in hue than the corolla, but in certain of the Stapelieae the effect of the double corona is much more remarkable than this. Here the outer lobes, next the corolla, may have much the flat oblong shape of anthers, and the inner lobes, which are more or less erect, may be slender and bent outwards much in the manner of styles. In certain species (see list of illustrations on p. 240 above) the result is that the centre of the flower is almost exactly like that of some more normal one.

Put concisely the problem here is whether or not it is reasonable to attribute this resemblance of one set of structures to quite a different one solely to chance, using that word in the sense of coincidence, coupled, perhaps with the inevitable central position of any corona. Difficult as it is to deny this explanation (if such a word is justifiable) it certainly cannot be considered altogether satisfactory. In such circumstances there is no reason why, for example, there should not be coronas in which the outer lobes were style-like and the inner lobes anther-like, in which case there would be no crucial resemblance, but there are none such recorded. Again the degree of coincidence required to produce such coronas as have been described can only be regarded as extreme. If it is not coincidence then presumably there must be some factor disposing the corona towards this form and that is even more difficult to understand. It may be that there is some innate tendency, in given conditions, for certain forms to appear in certain positions: it may be that there exists a tendency towards the formal (though not functional) replacement or restoration of lost or altered structures: it may be that the potentiality of shape in certain structures in the centre of the flower is, in fact, limited in some way, and has only a small range of expression in some respects. All these are among the interesting speculations which these coronas invite, but in the present state of our knowledge they can, unfortunately, be no more than speculations.

The second wide expression of false structures in the Asclepiads concerns the stigma-head. Usually, as might be expected in view of other circumstances, this is flat above and does not project beyond the stamens, so that it has no appreciable resemblance to any ordinary style and stigma, but in quite a notable proportion of the group the top of the stigma-head is prolonged and, often, divided, so as to reproduce, in varying degrees, the appearance of many normal styles and stigmas. There has, however, been no suggestion that these prolongations are functional in pollination and they are regarded as emergences to which such names as "styliform appendages" have been applied.

These remarkable structures pose the same problems as did the stapeliad coronas, but with two interesting differences. In the Asclepiads the corona is a normal and almost invariable constituent of the flower, and has, as it were, to be accommodated in any of its designs, but the prolonged stigma-head is a comparative rarity which contributes nothing to the essential plan of the flower and thus adds, rather than removes, a difficulty. Second, the apparent replacement or restoration here is not that of one structure by a different one but what can only be described as the occurrence of false versions of genuine tissues. The apparent stamens and styles in the Stapelieae case are made up of coronal tissues, but the prolongations and styliform appendages are made up of stylar tissue.

This being so the possible explanations are a little different. Assuming that the structures in question are, as is generally accepted, functionless (and were they not the whole design of the flowers and their pollinial organization would be meaningless) there are two leading possibilities, one that they are the relics of true styles, from which, in the course of time, the stigmatic function has migrated to the under side of the stigma-head, and the other that they are subsequent elaborations of flat stigma-heads. The former finds some support in the form of many apocynad styles and stigmas but involves an explanation of why, if these are indeed the circumstances, the relic structures should persist in only a few scattered asclepiads. The latter raises the difficult question of why these structures should appear at all, and here much that was said about the stapeliad coronas again applies, except that the restoration of a lost form by the tissues involved in its disappearance seems, at first sight, even less easily explicable, because this explanation would predicate that the disappearance of a structure in the course of evolution might be followed by its redevelopment in a functionless condition from the original source. Here the problem must be left but it may be pointed out that the circumstances just adumbrated are of special theoretical interest in relation to some of the facts described and considered in Chapter 10.

The Problem of Taxonomic Dichotomy

The last matter to be commented upon in this chapter is essentially different from the rest because it involves man-made classification, but it is one which is to be observed so widely in the Flowering Plants that it is highly desirable, by a brief discussion of the exceptionally good example of it afforded by the Apocynales, to consider what, if any, real significance may attach to it. It is the question whether there is any reality in the repeated dichotomies which, as was seen in Chapter 7, are so prominent a feature of the classification of that order, and, incidentally, also of so many others. Do the plants really fall into a binary arrangement of this kind which reflects a corresponding process in their evolutionary development, or is it merely that human analysis has a tendency to take this form?

It is a familiar fact that any collection of differing entities can be divided into two series and two only if the criteria used are a positive and negative pair. Thus all flowering plants could be divided into two assemblies according to whether their flowers are "white" or "not white", and characters of this sort are commonly used in simple botany books to help in the identification of plants, that is to say in the separation of one sort from another. But this is a process of elimination, rather than a classification in the true sense, because true classification has as its aims the isolation of groups of reality and, in the case of biology, the reconciliation of resemblance and kinship, taking into account for these purposes totality of similarity and difference in positive terms. There appears to be no theoretical reason why positive characters should be binary, and therefore no *prima facie* reason why the application of these characters should result in a classification exhibiting conspicuous dichotomy.

Yet the internal classification of the Apocynales is very notably dichotomous. The order itself consists of two families, or to use a less equivocal term two equivalent primary groups; each of these in turn reveals two obvious subgroups. Within the Cynanchoideae there are two series, one with pendent and one with erect pollinia; in the latter the pollinia are either single or double; and in those where they are single there are either anther appendages or not; and where these are absent there may or may not be hyaline edge to the pollinia; and so on.

Now it is obviously possible to arrive at a dichotomous or binary classification even when positive characters only are used provided that these are selected for their alternativeness rather than anything else, and in biological classification much may also depend upon the relative status conferred on the various groups. But it must be conceded that neither of these conditions quite applies to the case under discussion, where nearly all the facts mentioned relate to the possession by the plants concerned of characters which are almost entirely peculiar to these groups, as for instance the possession of pollinia, the consequent peculiarities of the stamens and the like. True, separation on these grounds may not reflect the truly "natural" classification and constitution of the group, which we might find, could we but discover it, to be more than dichotomous, but we can do no more than consider the facts as we know them and these certainly suggest the possibility, if nothing more, that there are true factual dichotomies in the Apocynales. The theoretical importance of this in relation to the nature of evolutionary processes is clear enough. If evolution has proceeded chiefly or entirely by the gradual accumulation of small differences there is no reason to expect that dichotomy will reveal itself to any conspicuous extent, but if on the other hand evolution has been predominantly a matter of large-scale mutation then the discontinuities which are the bases of dichotomies would be its almost inevitable concomitants, as, indeed, Willis[*] suggests in his theory of divergent dichotomous mutation.

[*] See Willis, J. C., *The Birth and Spread of Plants*, Geneva, 1949.

CHAPTER NINE

FLORAL AGGREGATION AND THE PSEUDANTHIUM

Simple Aggregation

THE astonishing multifariousness of the Flowering Plants may be ascribed largely to three circumstances—first, that every plant comprises a large number of distinct morphological characters or components; second, that many of these show, over the phylum as a whole, wide variation gamuts or ranges of values; and third, that there is little consistent correlation between the values of different characters. Of the many characters which show conspicuous variation gamuts, those of the size, number, and degree of aggregation of the flowers are not only obvious but also closely connected, since it appears to be the general principle that the smaller the flowers the more numerous and closely set they are. At one end of the sequence there is the extreme of the truly single- and large-flowered plant, and at the other is the plant which has great numbers of minute flowers so closely packed together that they are at least in contact, and sometimes actually fused.

Size and Number of Flowers

It is especially the degree of aggregation which is the subject of this chapter, but before embarking on this theme it is useful to make a few comments on the other two related components, *size* and *number* of flowers. Because of the accretive or accumulative growth plan of flowering plants the number of flowers in the individual is, in the last resort, governed by the size of that individual, but this is not the only correlation, and speaking generally it may be said that, among plants similar in size, the smaller the flowers the more numerous they are. There seems, however, to be a practical limit here in both directions so that in fact very few plants have, on the one hand, flowers more than a few inches across, or, on the other hand, flowers of microscopic dimensions. As regards prevalence also, the extreme one-flowered state is very uncommon, and virtually confined to certain geophilous monocotyledons (see p. 82), though of course many small plants, especially monocarpic annuals may have only a single flower if growing under conditions of semi-starvation.

On the other hand, since flowering plants may, within comparatively broad limits, reach almost any stature and since their flowers may be very tiny, one individual may bear enormous numbers of them. There must be many species in which the average number of flowers on the plants is to be measured in thousands, and in some, such as large willow trees, the numbers may well reach millions. At the same time, and for similar reasons, there may be a greater number of larger flowers on bigger plants than of smaller flowers on lesser plants, but this is perhaps stretching logicality too far, and by and large the number of flowers is in inverse proportion to their average size. The most multiflorous plants will therefore tend to be those in which are combined the greatest stature and the least flower-size.

There is also the relative flower-bearing capacity of plants to be taken into account and this varies very much. Some plants, as for instance the box, are scarcely more conspicuous when in bloom than at other times, but some, such as heather, flower so copiously as almost to conceal the vegetative parts. Still others, such as the cherries, flower profusely before the leaves unfold. One aspect of both these latter, however, is the concentration of flowering in time, and it may be doubted whether the actual number of flowers is really greater than in some plants which flower less strikingly but over a much longer period.

Although there are plenty of examples to the contrary it is usually the case that very small flowers are of simplified design, and this may well be traceable to one of the fundamentals of biological organization, cell structure, for it is a basic feature of this that however much whole individuals, or even their constituent tissues, may vary in size, the component cells do not do so in any comparable degree. In consequence the size of an organism is a rough measure of the number of cells it contains, and, also in consequence, an organ is unlikely to be below a certain size if it comprises more than a certain number of cells. This is of special potential significance in the andrecium of the flower which produces the pollen grains, each of which is, at one stage, a single cell. So long as a multiplicity of these grains is requisite there is a practical minimum for anther size and, so long as ordinary considerations of correlation prevail, of associated structures. It is therefore not surprising to find that small flowers are usually oligomerous, often lack all or part of the perianth, and may be unisexual. There is some indication also of the opposite state of affairs, namely that flowers with large numbers of parts are above the average in size.

Degrees of Aggregation

It is familiar that almost every degree of floral aggregation occurs in the Flowering Plants, though not with equal frequency, and the larger constituent groups of these, such as the families, may be regarded as of five kinds on this basis of comparison. These are:

1. Families which show a particularly wide variation of inflorescence form and which therefore illustrate the aggregation gamut in the widest fashion, these being especially noticeable in the Dicotyledons and well exemplified by the Rosaceae and Rubiaceae.

2. Families showing little or no floral aggregation, among which the Magnoliaceae and the Nymphaeaceae are perhaps the most noteworthy.

3. Certainly most numerous are families in which a partial degree of aggregation is general, or to be more precise, in which the less condensed forms of multiflorous inflorescences, such as racemes, panicles and cymes, are the rule, as in Crassulaceae, Caryophyllaceae, Ericaceae, Palmae and many other of the more general run of flowering plants.

4. Families in which there is a high degree of aggregation, with such condensed inflorescences as heads, spikes, and whorls of various kinds common, and often almost invariable. In these the flowers are more often than not sessile and in actual mutual contact, as for instance in the Labiatae, Plantaginaceae, Cyperaceae, Gramineae and so on. Sometimes there is actual fusion between adjacent flowers, and this is usually referred to as synanthy if the flowers, as such, are involved, and as syncarpy if it is a matter of the fruits, as in the pineapple. True synanthy is rather hard to define because when, as not infrequently happens, flowers are sunk in the axis which bears them, they are in complete continuity though not in actual contact. Again, fusion is

often a matter of the gynecium only and here the distinction between synanthy and syncarpy wears thin. The Rubiaceae in particular afford interesting examples of these points, among them *Pomax*, and the families Hamamelidaceae, Pandanaceae and Platanaceae are also noteworthy in this connection.

5. It might well be concluded that aggregation could scarcely go beyond this point, and in the strictly physical sense this is true, but it is one of the most remarkable features of the Flowering Plants that there is not uncommonly an even more complete and intimate association of the individual flowers. In this last extreme state, there is, over and beyond any mere aggregation of the flowers, such a measure of arrangement, organization and division of labour, both between the flowers themselves and between them and their immediate surroundings, that the inflorescence as a whole attains a degree of unity of design and function greater than any that may result simply from the proximity or contiguity of units all like one another. In its more intense expressions it is the essential nature of such an inflorescence that it does to a recognizable degree, function as and simulate a single true flower. It is an inflorescence of such a kind that the pollination of all its constituent flowers can be effected by the same sort of operation and with the same deftness and economy of effort as suffices for the pollination of an ordinary single flower of comparable size. An inflorescence of which these two statements are broadly true in combination is called a *pseudanthium* or false flower.

Since this definition of the pseudanthium contains elements of morphological degree and also of human appreciation it is inevitably somewhat subjective and, in many instances, it must be largely a matter of opinion as to whether the use of this term is justifiable or not, and this makes for some difficulty in presenting the relevant facts. However, it is clear that some plants deserve it less and others more, and if, as is most convenient on general grounds, the material is arranged roughly in this order, putting first what may be called subpseudanthial states and then the more proper pseudanthia, it is really of little consequence where exactly the arbitrary line between the two is drawn. This procedure will, moreover, not only emphasize what a very complete series of conditions exist but also will throw into even stronger relief the extraordinary nature of some of the more extreme circumstances.

Subpseudanthial States

By subpseudanthial is here meant any condition in which, quite apart from mere floral aggregation, there is some associated feature which, if intensified would tend to result in a more definite pseudanthium. In this sense subpseudanthial plants may also be thought of as exhibiting incipient pseudanthy.

The "target" type

In this, radial differentiation between the individual flowers of multiflorous inflorescences serves to enhance the target-like effect

FIG. 127. *Iberis gibraltarica:* plan of inflorescence, × 1½

FIG. 128. *Scabiosa pterocephalus:* plan of inflorescence, about natural size

of the whole and thus to give it more the aspect of a single floral unit. Sometimes, as in species of *Hydrangea* and *Viburnum*, it is merely the outermost flowers that are larger and more conspicuous than the rest; but in other cases there is some gradation of flower size from the centre outward and the outer flowers are increasingly zygomorphic. This produces the effect of a more solid centre surrounded by a periphery of petaline rays, and is perhaps best seen in such umbelliferous genera as *Orlaya*. It is also noticeable in some Crucifers where the racemes are strongly corymbose, as in the familiar *Iberis gibraltarica* (fig. 127). Here the lower (outer) flowers are appreciably irregular and are closely overlapped by the inner, the resulting target effect being most marked during the early stages of flowering while the upper (central)

flowers are still in bud and before the corymb has grown out.
There is a similar effect in certain capitate inflorescences such as
those of *Scabiosa pterocephalus* (*Pterocephalus parnassi*) (fig. 128),
but here the emphasis is rather on the close aggregation of flowers
rather than on differences between them, although even here the
peripheral flowers are usually larger and slightly irregular.

The "petaline" or "bracteate" type

In a second, and more heterogeneous, theme of subpseudanthy
the visual effect of the inflorescence is enhanced by the presence of
white or coloured bracts or analogous structures. The essential
element in this is that foliar structures which in other plants are
green here are white or coloured or otherwise peculiar (and often
enlarged also) so that they appear more or less like parts of a
corolla, and there are broadly two conditions of this. In what may
be called apical bracteation the uppermost leaves or axile bracts are
specially conspicuous, while in what may be called floral bracteation one, or more, of the floral bracts or even of the calyx lobes, is
unusual. The former is well seen in some Labiates (though it
occurs in various other groups as well) and is familar in *Salvia*, but
it is in such a plant as *Monarda didyma* (fig. 129) that, in conjunction with certain other organizational details, it contributes to the
highest degree of subpseudanthy. Here the inflorescences are most
often entirely apical and at the base of each there are two series of
coloured bracts, one immediately below the flowers, of narrow
pointed divergent bracts, and another below again of larger and
more pendent leafy bracts. Moreover the flowers in the inflorescences are so arranged, or at least their succession is so ordered,
that they open in successive whorls from above downwards. The
effect is that the flowering stem is terminated by a ring of open
flowers standing horizontal or more or less erect, within which are
the empty calyces of earlier whorls, while *below* are whorls of buds.
Visually the open flowers together simulate a single corolla or the
rays of a capitulum and the zone of narrow bracts simulates a
single calyx or involucre.

The second sort of bracteation is especially well seen in, though
not entirely confined to, the Rubiaceae, in which there is a remarkable series of conditions, exhibiting three main expressions. First,
are the not infrequent cases, including the genera *Kadua*, *Jackia*,

Manettia, Schizostigma, Pentaloncha, Dictyandra and *Alberta,* where some or all of the calyx lobes in every flower are enlarged, and either more or less coloured or at least unlike ordinary sepals. Second, there is the intensification of this state to that where one or more of the calyx lobes of one or more flowers in the inflorescence may be very much enlarged and coloured, so that they serve to draw immediate attention to the flowers which are often not much aggregated. It seems that there are three variations of this theme, namely, the enlargement of only one calyx lobe of one flower *per* inflorescence, as is described for *Warscewiczia*; the enlargement of one calyx lobe in each of a small proportion of the flowers, as in *Pogonopus, Pallasia, Schizocalyx, Calycophyllum, Capirona* and, best known of all, *Mussaenda* (fig. 130); and the enlargement of from one to three of the calyx lobes in many or perhaps all the flowers as in *Pinckneya* and *Cruckshankia*. Finally, in *Acranthera, Hymenopogon* and *Didymochlamys,* the inflorescence, which is terminal, is subtended by two or more much enlarged and usually white bracts. Apart from all these there are various other more special instances of bracteation in other families which may be exemplified in general by the genera *Bougainvillea* in Nyctaginaceae and *Schizophragma* in Hydrangeaceae, but something more must be said of four of them.

In some species of *Tacca,* of the monocotyledonous family Taccaceae, there are umbellate inflorescences at the tops of long stalks, and fairly ordinary individual flowers, but these have peculiarly modified bracts. Two of these are much enlarged and somewhat coloured and stand erect behind the flowers, while groups of others on either side of the flowers are thread-like and project laterally. The result is a highly organized unity in some ways distinctly comparable with that in *Dalechampia,* referred to below, but, as in that plant also, not such as to give a close resemblance to any particular euanthial flower.

In *Cyclanthus,* belonging to the Cyclanthaceae, there are coloured subtending bracts at the base of the highly concentrated and almost cylindrical inflorescence, but there is not much close resemblance to any true flower partly because the bracts are rather unordered, and partly because the innumerable small unisexual flowers are arranged on the inflorescence axis in alternate whorls of males and females.

In *Davidia*, of Nyssaceae (fig. 131), the so-called "handkerchief-tree," the small flowers of the inflorescence are aggregated into a spherical button-like head in which one flower is female and the rest male, and this is subtended by two large, lax, white, unequal bracts.

In *Dalechampia* (figs. 132, 133), which constitutes a tribe of the Euphorbiaceae, there are about a hundred species, found mostly in America but also in Africa, Madagascar and India, and in the most characteristic of these each inflorescence is compound and on a long axillary stalk, and is subtended by a pair of broad coloured leafy bracts. The compound inflorescence itself consists of a lower 3-flowered female dichasium; an upper male pleiocasium, in which there are three 3-flowered and two 1-flowered inflorescences, each subtended by its own bracts; and of a number of sterile flowers united into a lobed and crested mass. To paraphrase this astonishing description it may be said that there are three distinct kinds of flowers, each in its own sort of partial inflorescence, that two of these are bracteate, and that the whole is subtended by two much larger bracts, giving the whole a remarkable and quite strongly sub-pseudanthial effect. The whole may be called a subpseudanthium or pseudanthium (according to taste) composed of partial inflorescences, and made all the more complex by the monoecism of these entities. On the assumption that this interpretation is correct *Dalechampia* must surely rank as one of the most extraordinary of the Flowering Plants.

The "fig" or "mulberry" type

In another kind of subpseudanthy the theme is mainly the exceptionally close aggregation of small flowers, though it is usually accompanied by some degree of organization, and such aggregation may amount to actual continuity between the flowers. This state of extreme floral aggregation is particularly well illustrated by the two families Moraceae and Urticaceae, both with very small and simple flowers, usually unisexual, and since the conditions here are relatively unfamiliar they may be mentioned in some detail.

In the Moraceae these conditions culminate either in large fleshy compound fruits or in expanded receptacular inflorescences and false fruits. The Fatoueae have inflorescences which are not remarkable, but elsewhere there is at least some evidence of

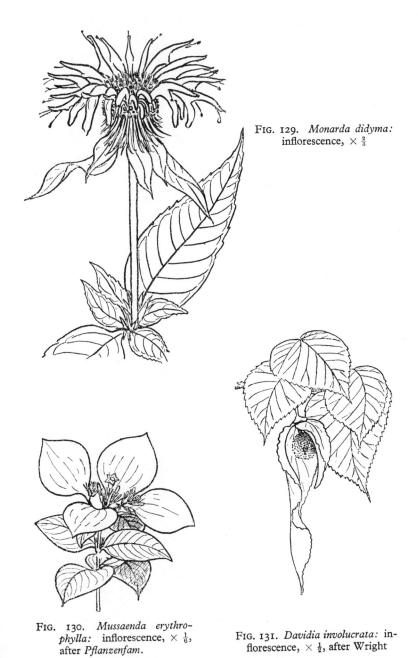

FIG. 129. *Monarda didyma:* inflorescence, × $\frac{2}{3}$

FIG. 130. *Mussaenda erythrophylla:* inflorescence, × $\frac{1}{6}$, after *Pflanzenfam.*

FIG. 131. *Davidia involucrata:* inflorescence, × $\frac{1}{2}$, after Wright

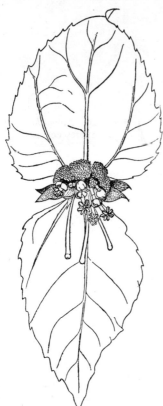

FIG. 133. *Dalechampia roezliana:* vertical section of inflorescence, × 2

FIG. 132. *Dalechampia roezliana:* elevation of inflorescence, × 2

FIG. 134. *Antiaris toxicaria:* showing inflorescences, × ¼, after *Pflanzenfam.*

FIG. 135. *Rhodoleia championii:* showing inflorescences, × ½, after *Pflanzenfam.*

FIG. 136. *Diplolaena grandiflora:* showing inflorescences, somewhat reduced, after *Pflanzenfam.*

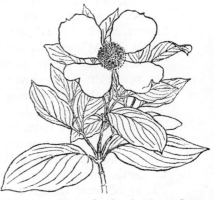

FIG. 137. *Darwinia macrostegia:* showing inflorescences, somewhat reduced, after *Pflanzenfam.*

FIG. 138. *Cornus florida:* showing inflorescence, × ½, after Baillon

extreme aggregation, and in the Moreae this takes the form of multiple fruits in which the ripening flowers, with their fleshy perianths, are almost united. In *Perebea* and *Helicostylis* there is both synanthy and syncarpy. In *Maclura, Chlorophora* and *Broussonetia* there is syncarpy, the first-named having a yellow compound fruit, and the second synanthy in the male as well. In *Plecospermum* the individual female flowers are widely spaced and sunk in a spherical fleshy mass which later becomes a compound fruit. *Cudrania* and other genera also show syncarpy, culminating in the genus *Artocarpus* (fig. 76), where the large compound fruits, rich in starch, are known as bread-fruits. In *Castilloa* and *Pseudolmedia* the flowers are on flattened unisexual receptacles which are not noticeably bracteate, but in *Antiaris* (fig. 134) the male flowers are on flat round receptacles furnished with a rim of small bracts. In *Phyllochlamys* also the small sessile clusters of male flowers are subtended by a series of bracts. In *Dorstenia* (fig. 50) the male and female flowers are intermixed and are borne more or less immersed on fleshy green receptacles of very varied and often strange outline, and even, in some, strongly radiate. The arrangement of the sexes is also often noteworthy. In *Trymatococcus* the receptacle is concave with male flowers towards the periphery and one female flower towards the centre; in *Brosimum* there are more or less spherical receptacles on which all except the one central flower are male; in *Scyphosyce* there is further differentiation; and in *Lanessania* the receptacle is turbinate and covered with bracts, bearing on its flat upper surface many more or less immersed male flowers surrounding one single deeply sunken central female flower. Finally, there is the very special condition of the great genus *Ficus* (fig. 35), with its deeply invaginated receptacular inflorescences or figs, and its satellite *Sparattosyce* from New Caledonia in which the deeply invaginated condition persists only during flowering, the receptacle opening out as the fruits ripen. It is of some interest to note that several members of the family Monimiaceae also have states not unlike some of those just described, and also that in *Pterisanthes*, of the Vitaceae, an inflorescence superficially not unlike that of *Dorstenia* occurs.

In the rather smaller family Urticaceae the same sort of series is to be seen, though less accentuated. Thus, in most of the Urereae, which include the common stinging-nettles, there is little peculiarity, but elsewhere, in such genera as *Lecanthus* and *Procris* the flowers are on flat receptacles, while in *Elatostema* the receptacles are involucrate. In the Boehmerieae there is commonly a close aggregation of the flowers, especially of the females; in the Parietarieae the inflorescences are subtended by bracts; in the Forskohleae there are two remarkable genera *Forskohlea* and *Droguetia*. In the former there are small turbinate receptacles bearing both male and female flowers, or one of these only, and surrounded by from four to six relatively large perianth-like bracts, and the latter is similar except that the involucre of bracts is cupular and not lobed.

Conditions parallel to some of the above occur in the Euphorbiaceae which also have simple flowers. In *Uapaca* the heads of small male flowers are subtended by a ring of involucral bracts, and in the genus *Pera* the generally dioecious flowers are enclosed in almost spherical involucres of very few parts. Finally, there is the remarkable state of *Dalechampia* already described.

The "involucrate" type

Several of the plants just mentioned serve to introduce this type, and call attention to what is in fact by far the most important single ingredient in the development of pseudanthy, the subtention of the flowers of the inflorescence, whether completely aggregated or not, by a series of bracts. No other factor gives such unification to the contents of an inflorescence as a whole, and it is no exaggeration to say that the condition of the bracts in these cases is the best single criterion for the use of the word pseudanthium. If they form a single well-defined, regular, spreading or multiseriate whole a true pseudanthium almost inevitably results, even if, as in some of the above, the flowers themselves are very inconspicuous.

Of this involucration, as it may be called, one or two of the least marked states may first be noticed. In *Dais* (Thymelaeaceae) the flowers are in small capitate groups subtended by a simple series of a few bracts at the tops of long peduncles. These bracts are not highly organized, but more important is the fact that the flowers are long-tubular and much exceed them, so that they have little more visual value than the bracteoles at the base of some pedicellate umbels. In some species of *Lasiosiphon*, also Thymelaeaceae, the involucration is a little more marked. A comparable, though rather different, degree is illustrated by species of *Acrocephalus*

(Labiatae) where the globular flower masses are subtended by flat bracts, or leaves, which differ somewhat in shape from the true foliage leaves. Rather more definition is to be seen in the three like genera *Sphenodesma*, *Congea* and *Symphorema* (Verbenaceae), though here the effect is somewhat incomplete because the bracts which subtend the flowers are narrowed towards the base and so fail to form a real cup. Another emphasis is that in which the bracts are relatively inconsiderable but the flowers small and very regularly set, as in the teasel, *Dipsacus*, of Dipsacaceae, and the sea-hollies, *Eryngium*, in Umbelliferae, especially such species as *E. humile*, *E. longipetiolatum* and *E. rosei*. The genus *Gomphrena* (Amaranthaceae), some species of which, e.g. *G. macrocephala*, are almost fully pseudanthial, also calls for note here.

A condition not unlike that of *Dais* is seen in *Loranthus lageniferus* in Loranthaceae, where the flowers are in small clusters, each in a deep campanulate involucral cup, but elsewhere in this same family the genus *Lepeostegeres* shows quite a different state of affairs, the small and simple flowers being seated at the bottom of an urn-shaped receptacle formed of decussate bracts, through the aperture of which the styles project, much as they do in the female inflorescence of *Corylus*.

Stages of involucration are well illustrated by the Balanophoraceae and Umbelliferae. In the former family *Thonningia*, for instance, shows aggregates of unisexual flowers subtended by a long series of bracts, but these are scarcely condensed and overlapping at all; in *Langsdorffia hypogaea*, also, this is true of the male, but in the female inflorescence the bracts form a typical involucre, so that the whole may almost be called a capitulum; in *Scybalium fungiforme* both male and female inflorescences have well-defined capitula quite comparable with those of Eriocaulaceae or even some of the Composites. In the Umbelliferae there is an interesting series in the genera *Oliveria*, *Hacquetia*, *Lagoecia*, *Bupleurum*, *Astrantia* (fig. 139) and *Xanthosia*. In *Oliveria* the umbels are more or less as usual in the family, but the bracteoles subtending them form a continuous ring and are the same length as the pedicels, so that the flowers seem to be in a cup. In *Bupleurum* the flowers are more or less sessile and set off by a few relatively dull-coloured bracts, but the general level of organization is fairly high, especially in such species as *B. stellatum* and

B. ranunculoides. In *Astrantia* there is some sexual dimorphy in the shortly stalked flowers, and the bracts are many and often more highly coloured, so that the whole effect is enhanced. In the Australian genus *Xanthosia* the umbels are subtended by particularly petal-like bracts and some remarkable effects result.

There is an interesting series of types also in the Bromeliaceae. In quite a number of species belonging, especially, to the genus *Nidularium* the close, sessile groups of flowers are sunk deep in the

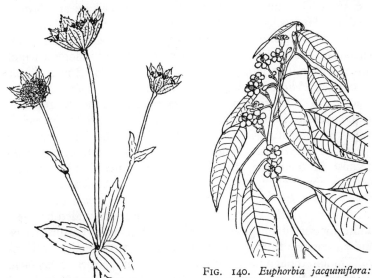

FIG. 139. *Astrantia major:* inflorescences, × ½

FIG. 140. *Euphorbia jacquiniflora:* showing the flower-like inflorescences (cyathia), × ½, after *Bot. Mag.*

centre of the leaf rosette, so that the uppermost leaves subtend them by a kind of involucre; but in some species, among them *Aregelia carolinae*, these latter may be partly or wholly brightly coloured so that they simulate a perianth. Sometimes when this happens, as in the species noted, the innermost leaves are wholly coloured, but the more outer (and larger) ones only basally so, so that the visual radiating colour patches are all more or less the same size. Elsewhere in the family, as in *Guzmania cardinalis* and *Aechmea gigas*, there are long-stalked flower groups subtended by coloured bracts, which are notably subpseudanthial, a condition which is specially well seen in *Aechmea viridis*. This last takes us easily to the remarkable genus *Phaeomeria* in Zingiberaceae, where *P. magnifica* in particular with orange flowers set in a reddish involucre, is almost fully pseudanthial, and *P. hemisphaerica* is but little less so.

But even these do not exhaust the subject of involucration, and some kind and degree of it is to be found in many families. They are much in evidence in the Rubiaceae, where the best examples are perhaps *Geophila involucrata*, *Stipularia africana* and *Thiersia insignis*, while among others worth noting elsewhere are *Boottia cordata* (Hydrocharitaceae); *Archytaea multiflora* (Theaceae); species of *Eriogonum* (Polygonaceae); *Cornucopiae* (Gramineae); *Hernandia* (Hernandiaceae) in which each partial inflorescence of three flowers, one female and two male, is enclosed in a simple involucre; *Peltodon* (Labiatae); *Ptilotus helichrysoides* (Amaranthaceae), in which the uppermost leaves beneath the inflorescence are compressed into a petaline whorl; *Staavia* (Bruniaceae) which is somewhat similar; *Leianthus umbellatus* (Gentianaceae); and some species of *Freycinetia* (Pandanaceae). Also interesting, especially because of their size, are the huge bracteate, though scarcely perhaps involucrate, heads of *Brownea* (Papilionaceae) and of *Dombeya wallichii* (Sterculiaceae). In *Androcymbium* (Liliaceae) the flowers, which are in a simple umbel, are surrounded by a few of the uppermost foliage leaves which are different in posture, shape and appearance from those below. Finally in connection with involucration, though not an example of it, some species of *Neptunia* (Papilionaceae) are very notable because the flowers, which are closely aggregated, are segregated into hermaphrodite above and female below, which is, of course, the condition most characteristic of the Compositae, the family in which an involucre is particularly developed (see next chapter).

So much for some of the leading subpseudanthial states, among which four types have been distinguished—the "target", the "petaline" or "bracteate", the "fig" or "mulberry", and the "involucrate". We must now pass on to what in comparison with these, may be called pseudanthia proper, and it will be seen that in nearly all of them it is the intensification of one, or more than one, of the principles embodied in these four types which is the chief factor in their organization.

Pseudanthial States

Miscellaneous

Nearly all the real pseudanthial types are represented by large assemblies of species receiving high taxonomic rank, but there are a few others, all single genera or parts of genera, and these include some of the most interesting and striking of all. In *Rhodoleia* (Hamamelidaceae) (fig. 135) and in some species of *Diplolaena* (Rutaceae) (fig. 136) the inflorescences (which are widely spaced) and the surrounding bracts are of such size, number, posture and general appearance, that each looks very like the single flowers of such a genus as *Camellia*. In other species of *Diplolaena* the bracts are smaller and the pseudanthium more like a polystemonous flower with prominent stamens. An even more extreme condition of this is seen in *Sycopsis sinensis* (Hamamelidaceae) where the bracts are so small as hardly to constitute an involucre. In some species of the genus *Darwinia* (Myrtaceae), e.g. *D. fimbriata* and *D. macrostegia* (fig. 137); in *Cavendishia cordifolia* (Ericaceae); and in *Pimelea physodes* (Thymelaeaceae), the pseudanthia are more or less pendent, but otherwise much the same.

Rather different is the state in which there is a simpler corolline appearance, chiefly because the petaline bracts are fewer and more spreading. Such is seen in *Houttuynia* (Saururaceae), in which a few white bracts subtend an elongated spike-like true inflorescence, but the best examples are in the genus *Cornus* (Cornaceae). In some species here the flowers are quite sessile in a small dense cluster which is set off by a whorl of relatively large white or pink bracts. *Cornus suecica* is perhaps the most generally familiar, but the larger, woody species *C. nuttallii* and *C. florida* (fig. 138) are much more striking, with pseudanthia up to three inches across. *Parrotiopsis jacquemontiana*, another hamamelidaceous plant, is not unlike these, but the true flowers are rather more conspicuous and the bracts less so.

The Eriocaulaceae (the Pipeworts)

This is perhaps the least highly organized of the large pseudanthial groups (fig. 10) but among monocotyledons affords a remarkable counterpart to the Composites. The Pipeworts, as these plants are called, are rather small with some vegetative resemblance to the Plantagos, but their very inconspicuous and generally monoecious flowers are massed in flat or convex heads subtended by involucres of bracts (e.g. *Mesanthemum prescottianum, Eriocaulon stellare, Paepalanthus lancifolius, P. xeranthemoides* and *P. niveus*) and arranged in compound inflorescences such as cymes and umbels. Neither the bracts nor the flowers themselves, however, are very brightly coloured.

The pseudanthial Proteaceae (the Proteads)

The family Proteaceae comprises two subfamilies, the Persoon-ioideae and the Grevilleoideae, the former being characteristic of Africa, though it is also well represented in Australia, and the latter being characteristic of Australia and absent from Africa. In general terms the family exhibits two kinds of inflorescence, one with a fairly high, but by no means extreme, degree of simple floral aggregation, and the other a highly developed pseudanthial state. Also generally speaking the simpler of these types is characteristic of Australia, while the pseudanthia are more or less confined to Africa. These pseudanthial Proteaceae are one of the most striking features of the flora of South Africa, and are remarkable for their size and multiplicity of parts, especially in the genera *Protea* (fig. 49), *Leucadendron* and *Leucospermum*. The flowers, which are homochlamydeous and often more or less irregular, are closely packed together and surrounded by a multiseriate perianth-like series of involucral bracts, and in some the organizational level is not particularly high. In others, however, the size, colour and ornamentation of the bracts, together with the colour and texture of the contained flowers produce pseudanthia of exceptional dimensions and substance.

The Euphorbieae (the Spurges)

In this subfamily of the Euphorbiaceae, of which the genus *Euphorbia* accounts for all but a small part, each inflorescence or partial inflorescence consists, essentially, of a cup of varying depth, within which are set a number of simple male flowers, generally if not always consisting each of a single stamen only, and in the centre a single long-stalked female flower, consisting of an ovary only, the whole being called a cyathium. The cup is usually regarded as being developed from an involucre of bracts and is itself normally green, but the pseudanthial effect is commonly heightened by the fact that it bears on its edge a series, most often five, of expanded, more brightly coloured, and sometimes petal-like glands. In extreme cases, such as that of *E. jacquinifloria* (fig. 140), or less so of *E. caput-medusae*, the mass effect of these cyathia is of an ordinary branched inflorescence of many small to medium, brightly coloured, pentamerous flowers, though in most species of temperate regions the glands are smaller. In *E. grandis* the glands are fringed with bi- or tri-furcate appendages, and the cyathia are subtended by broad, long-tailed bracts.

Sometimes the cyathia are subtended by coloured bracts of more leafy form. *Euphorbia punicea* is a simple example of this, but the most striking cases are in that section of the genus which is often separated as *Poinsettia*, where the cyathia are rather different in shape and are borne in clusters surrounded by large and handsome coloured leafy bracts. In all these the effect of bracteation is to produce a more or less definite *pseudanthium of pseudanthia*, a level of organization which will be encountered again in the next chapter. *Euphorbia heterophylla*, which has scarlet bract-like patches at the bases of some of the upper leaves, is also worth mention here because these patches have often a border of black which, simulating a shadow, makes them appear to stand away from the leaves which actually bear them.

Among the smaller, satellite, genera of Euphorbieae there are other interesting variants of the cyathium, though most of them result in something less rather than in something more pseudanthial. In *Anthostema* and *Dichostema* the flowers have a simple form of perianth; in *Diplocyathium* the involucral bracts or glands are in more than one ring; in *Synadenium* they are fused together; while in *Monadenium*, *Stenadenium* and *Pedilanthus* the cyathia are asymmetric.

The Aroids

The Araceae or arum-lilies (fig. 14) are mostly tropical, but the northern temperate members include one of the simplest, *Acorus*, as well as some of the best developed, such as *Arum* and *Arisaema*, and a comparison of these is interesting. *Acorus* is, like a few others of the family, actually not pseudanthial at all because its inflorescence is merely a short condensed spike of similar simple flowers without bracteation, but the two latter have well-developed pseudanthia. In them there is a spicate elongated floral axis as before but this is elaborated in a number of ways, chief of these being that the whole axis is subtended by a large bract called a spathe which contains the spike in a kind of oblique pocket. Further, the upper part of the spike is a sterile, and often coloured, club called the spadix. Nor is this all for the flowers themselves are usually achlamydeous, unisexual, arranged in separate zones, and sometimes aborted into hair-like structure. In pollination the whole structure is involved as a unit in a characteristic process and, therefore, well merits the name of pseudanthium, but it is very interesting to note that it does not in fact resemble any particular or even generalized kind of euanthial flower.

Not all the aroids conform to this particular pattern and several others call for comment. In some species of *Pothos* the spike is thin and lax and the individual flowers are slightly spaced from one another so that there is even less aggregation than in *Acorus*. In some species of *Anthurium*, among them the well-known greenhouse plant *A. scherzerianum*, the spike, though simple and covered with flowers of one sort only, may be twisted, and is subtended by a spreading, open brilliantly coloured spathe. In *Symplocarpus* the floral axis is swollen and largely hidden in the deeply concave spathe. In *Stylochiton* the female flowers are in a ring at the base of the floral axis, which is very thin and which bears rather scattered male flowers, and the whole pseudanthium is partly buried in the soil. In the Staurostigmateae the spadix may be partly or wholly adnate to the spathe, which is then not uncommonly convex rather than concave. Finally in certain aquatic members of the family, especially *Cryptocoryne* and *Pistia*, the typical condition, though still recognizable, is distinctly simplified.

Further, in certain genera there is a remarkable quantitative elaboration, well illustrated by two species of *Amorphophallus*. In *A. campanulatum* the pseudanthium has a full complement of parts, but the spathe is much augmented and frilled; the spadix is very stout and distinctly zoned; and the sterile apical part of it is greatly enlarged and irregular in shape. Added to this all the parts tend to be of lurid colours and the whole effect is most bizarre. In *A. titanum* there is less morphological elaboration but the whole pseudanthium may be several feet in height, making it and those of allied species by far the largest of all pseudanthia.

The Compositae (the Composites)
This great family, which provides the last of the major pseudanthial groups, is both quantitatively and qualitatively of so much more significance than any of the others that the whole of the next chapter is devoted to it, and it need only be mentioned here in order to complete the tale of pseudanthial plants.

NOTE ON LITERATURE

Further reference to the literature of the pseudanthium will be found in
Troll, W.: *Organisation und Gestalt im Bereich der Blute*, Monographien aus dem Gesamtgebiet der Wissenschaftlichen Botanik. Berlin, 1928.

This work is discussed at some length in
Arber, A. : *The Natural Philosophy of Plant Form*, Cambridge, 1950.

There may also be consulted
Arber, A.: "The interpretation of the flower : . . ." *Biol. Rev.* **12**, 1937.

CHAPTER TEN

THE COMPOSITAE

THE family Compositae is by a comfortable margin the largest, and one of the most completely distributed, in the Flowering Plants, and derives its name from the fact that, in all the more typical members, what appear to be the flowers are in fact separate inflorescences, each composed of a group of flowers or *florets* enclosed within an involucre of bracts (fig. 141). Also as a general rule these capitulate inflorescences are arranged in compound inflorescences, almost always of a cymose sort. Both visually and functionally the capitula play the part of the single flowers in most other families, and the Composites are therefore strictly and exclusively pseudanthial, and the most extensive instance of this kind of organization. Moreover, in many Composites there is a differentiation among the florets of the capitulum which gives a pseudanthial effect unmatched anywhere else in the Flowering Plants. This pseudanthy not only makes the family very homogeneous but is the chief reason why it is regarded as taxonomically isolated, for there is no other group (except the tiny family Calyceraceae) which, on the usual criteria, can be held to be closely parallel to the Composites and similarly pseudanthial.

Characterization of the Composites

It is not surprising to find a comparatively wide range of structure in so large a family, but this variety has mainly two sources, the vegetative organs and the form and arrangement of the florets in the capitulum. The homogeneity of the family arises from the invariable occurrence of one kind of primary inflorescence only, the capitulum, and from the characteristic design of the florets, so characteristic indeed, that no single Composite floret is likely easily to be mistaken for anything else even when divorced from the plant which bore it.

The reasons for this are twofold. First, there are to be seen in all normal Composite florets a combination of four more formal circumstances:

1. an inferior ovary of one loculus, containing one erect basal ovule, and ripening into a false achene.
2. the absence of a normal calyx and, commonly, its replacement by a pappus.
3. the presence of a corolla of ordinary coloration and texture.
4. epipetalous stamens with the anthers joined into a ring round the line of the single, generally more or less bilobed, style.

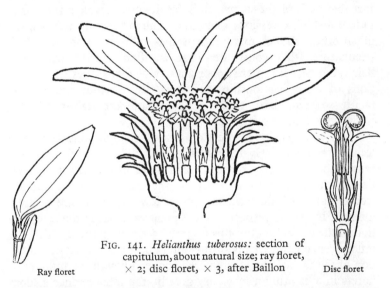

Ray floret

FIG. 141. *Helianthus tuberosus:* section of capitulum, about natural size; ray floret, × 2; disc floret, × 3, after Baillon

Disc floret

Second and less formally, though in fact even more significant, each of the structures contributing to the above four "characters" has what can only be described verbally as an appearance peculiar to the Composites. A descriptive phrase like one of those just used, such as "an inferior ovary of one loculus, containing etc." sounds at first adequate enough, and it is indeed as comprehensive as the average taxonomic statement, but it is actually so generalized that it would be quite impossible without either a living specimen or some previous experience to deduce from it exactly what the state of affairs in the Composites is. This is because the phrase inevitably enough goes only some of the way towards a complete and altogether unambiguous description. It may be that what our sight tells us is peculiar to the Composites, is something which cannot be put wholly into words, though it could certainly be described in greater detail than above, but this does not make it any the less real and ultimate. Much the same is true of any one of the other three characters, as, for instance, the anthers, which have a very special design, and this is why it is exactly true to say that while the Composites are, formally, distinguishable from all other plants in having, associated with the capitulum, this particular combination of floral characters, their real individuality springs from the fact that each of these characters has its own peculiar and partly indescribable expression. The same thing is true, in various ways, of all other families too, and it is the ability to appreciate these peculiarities that constitute what is often called "taxonomic flair", but here it is more important to realize that they are indications of close relationship, and it is because they are at least tacitly so recognized, that the Composites are rightly called a well-marked and isolated family.

The heterogeneity of the Composites, though considerable in total, is comparatively restricted in kind. The family contains a remarkable range of herbaceous plants and also some very specialized more robust forms, such as the giant Senecios, but the woody habit is little seen, though its few exemplars are of great interest. In certain respects there is great variation in the composition of the capitulum but this is circumscribed by the fact that the inflorescence is invariably of this kind and by the strict combination of the formal characters noted above. Though the number of florets in a capitulum may vary greatly, the inflorescence seldom ceases to be recognizable as a capitulum. Similarly though there is much variation in the shape of the corollas, this never completely overshadows the characteristic facies of the floret as a whole. There is a very wide range of form in the pappus, also, but since the pappus of the Composites is unrepresented elsewhere, this does not weaken the coherence of the family. Again, the distribution of sex among the florets is almost as varied as it can well be, yet it in no sense obscures the Composite nature of the florets. In short there is almost infinite variety in the Composites, but it is, to use a musical metaphor, compounded of the variations on a single and very precise theme.

Leading patterns in the Composites

Although there is such multiplicity in the Composites most members of the family conform to one or other of a small number of prevalent patterns, so that the family consists, essentially, of a few well-marked and familiar types, each represented by hundreds of species, and of a great number of other less prominent types each represented by only a handful of species or even by one only. On the whole it is true to say that the widest departures from the commoner expressions are among the latter, though this is a difficult matter to confirm because the commoner a condition is the less aberrant it appears to be.

The prevalent patterns may be said to number seven and may be summarized as a preliminary to their more detailed consideration, as follows:

1. The daisy pattern.
   Capitula with both peripheral ligulate (ray) florets, and central regular (disc) florets.
2. The everlasting-flower (*immortelle*) pattern.
   Capitula with disc florets only, but the bracts of the involucre (phyllaries) partly or entirely chaffy, coloured and ray-like.
3. The dandelion pattern.
   Capitula with ligulate florets only. Corollas generally yellow.
4. The thistle pattern.
   Capitula of disc florets only, the outer ones often slightly irregular, in a deep involucre. Corollas generally purplish.
5. The discoid (*Vernonia*) pattern.
   Capitula of disc florets only, medium in size, and usually arranged in open compound inflorescences.
6. The wormwood pattern.
   Capitula generally of disc florets only, but always small and rather closely aggregated.
7. The cudweed pattern.
   Capitula generally of disc florets only and so small and crowded together as to be separately indistinguishable without close examination.

1. *The daisy pattern* (*fig.* 142).

This is to be regarded as the typical pattern of the family in so far as it embodies the popular name for the members of the family. It may also be regarded as typical in the sense that it embodies the highest expression of the pseudanthial theme of the family, because it is the type with the greatest complexity and differentiation within the capitulum, a point of particular interest to us here.

The specialized nature of the Composite capitulum and the taxonomic isolation of the family has led to much speculation about its origin and history. It is not difficult to derive it, convincingly, from one or other of several simpler kinds of inflorescence, but it can be accounted for most simply by supposing it to be the result of the telescoping of a spicate raceme, and this is perhaps the most widely held theoretical opinion. A raceme is an indefinite inflorescence in which the flowers, more or less sessile if it is spicate, open in acropetal succession. In such inflorescences the individual flowers are normally each subtended by a floral bract, and beneath the lowest flowers there is usually a series of larger but empty bracts. If such a structure were telescoped we should therefore expect to find a circumferential zone of larger bracts surrounding a number of flowers of which the central ones will be the youngest, each flower being subtended by its own smaller bract. This is the general condition of affairs in the Composites and why a racemose origin has been so often suggested for the capitulum. Only, indeed, in the daisy pattern is there a really additional feature, the presence of two kinds of floret, ray and disc, and this may be accounted for on the supposition that telescoping has here been accompanied by floral dimorphism.

In this sense the floral dimorphy of the daisies is something over and above the usual specification for the family, and since these plants have all the more ordinary characters as well they must be regarded as the most highly differentiated of the Composites. Those daisies indeed, which possess receptacle floral bracts may, since they lack none of the features which capitula may commonly possess, and exhibit, to boot, structural and function dimorphy of the florets, be regarded as being, on the structural side, the most *complete* Composite type in existence today, and on the functional side as showing the most extreme condition of pseudanthy. In other words, and assuming the origin of Composites from plants

with simple homoflorous inflorescences, the daisies have under-
gone more change in the course of time than any of the other great
patterns of the family. The conception of this, which has been
called the *complete* Composite type, because it embodies all
components, is a very important one.

It is difficult to compute the proportion of the daisy type in the
family as a whole because the presence or absence of ray florets is
neither invariable in nature nor used consistently in classification,
and also because there may be small rays in capitula so small and
closely packed that their conspicuousness is to all intents and
purposes lost, but at all events daisy types appear in at least
25 subtribes, distributed over most of the tribes. Nevertheless a
rough computation based chiefly on the larger genera suggests
that they by no means form a majority of the family. Nor is the
type markedly restricted to any particular part of the world,
though they are perhaps more numerous outside the tropics.
They are rather specially emblematic of North America and
Mexico, and from this region come many of the species of the
genus which is most useful as an illustration of the "complete
Composite", *Helianthus*, containing the sunflowers (fig. 141).

2. *The everlasting-flower (immortelle) pattern (fig. 143).*

This finds an appropriate place next to the daisies because it does,
effectively, achieve the same function and pseudanthial end
though in a different way, the corolla-like ray florets of the daisies
being replaced by coloured members of the involucre. These are
almost invariably more robust and rigid than true petals and
therefore neither fade nor wither in the same way and this is why
the plants of this pattern have gained their more generally familiar
names.

Everlasting flowers are seen at their most typical in the great
genera *Helichrysum* and *Helipterum*, and the pattern is largely
confined to their section of the family, the Gnaphalinae. This
group is world-wide as a whole but its most typical members are
chiefly found in warm countries.

3. *The dandelion pattern (fig. 144).*

Many attempts have been made to divide Composites into two or
more smaller families, but they are so homogeneous that this has

FIG. 142. *Calendula officinalis:*
× ⅔, after Baillon

FIG. 143. *Helichrysum gulielmi:* × ½,
after *Bot. Mag.*

FIG. 144. *Leontodon autumnalis:* × ½,
after Fitch and Smith

FIG. 145. *Carduus crispus:* × ½,
after Baillon

FIG. 146. *Vernonia anthelminthica:* × ½,
after Baillon

FIG. 147. *Artemisia absinthium:* × ½

FIG. 148. *Gnaphalium uliginosum:* × ½

never become established practice. Nevertheless, nearly all systems of classification divide the family into two primary parts, the Liguliflorae, or plants of the dandelion pattern in which all the florets are ligulate (like the ray florets of daisies), and the rather weakly-named Tubuliflorae in which at least some of the florets in each capitulum are more or less tubular and regular.

The totally ligulate capitula of the dandelions make them, in one sense, no more different from the daisies than are the types in which all the florets are regular, but sundry small details help to mark them off more sharply, while, in addition, they have milky latex. In a few of the tribe Mutisieae of the Tubuliflorae there are capitula which contain none but ligulate florets (though these differ somewhat from those of the Liguliflorae), and in one or two of the Tubuliflorae, such as *Gundelia* and *Warionia*, there is milky juice. Again, *Scolymus* in the Liguliflorae has vegetative parts very like those of some thistles and conversely *Emilia*, in the Tubuliflorae, vegetative parts like those of *Sonchus*, while *Porophyllum* and a few others of the Tubuliflorae have characters of the involucre and other details of the capitulum very like those of some Liguliflorae. Other interesting, and perhaps associated, points about the dandelions are that they often show regular opening and closing of the capitula, and that the ligulate florets tend inevitably to overlap considerably.

The dandelion pattern is found all over the world but chiefly outside the tropics, and especially in the northern part of the Old World. It is very relevant to its evolutionary consideration that many of these plants show some kind of apogamous reproduction.

4. *The thistle pattern* (*fig.* 145).

This is not, in all its expressions, very sharply marked off from others, such as the everlasting-flowers and the Vernonias, but in its typical state it is most distinctive and conspicuous. The essence of the pattern is that the involucre is particularly highly organized and generally so deeply flask-shaped that the corollas project from it only because they are unusually elongated, and this gives to the capitulum as such, and irrespective of the florets, a remarkable degree of definity, so that it may with some justification be claimed that the thistle pattern is one of the most highly elaborated in the Compositae. As in the dandelions there are accessory or contributory characters which may or may not appear in combination, and chief among these are the familiar prickly or spinous vegetative parts; a tendency towards obliquity in the corollas, especially in those of the outer florets; the protrusion and prominence of the andrecia; and the purplish corolla colour. The second of these is a striking instance of how quite a minor feature may, when much emphasized, produce remarkable visual effects, as in the cornflower (*Centaurea cyanus*).

The tribe Cynareae to which the thistles belong contains a few other types as well and is widely, but very unevenly, distributed. Its members are specially characteristic of Europe, North Africa and West Asia, and there are very few in the southern hemisphere.

5. The discoid (Vernonia) pattern (fig. 146).

The characteristic of this pattern is that the capitula, which are comparable in size with those of the daisies or dandelions, contain only disc florets which are all alike both in shape and sex, and usually purplish or reddish in colour. These strictly homogamous capitula are, in terms of differentiation, and therefore perhaps of development, the simplest expressions of the Composite condition, and particular instances among them are of special interest and importance for this reason. This pattern is above all seen in the great genus *Vernonia*, which is very characteristic of the tropics, and secondarily in the larger-capitulate species of the great genus *Eupatorium*, most of which are American. It has a good deal in common with the thistles but in sharp contrast to them it is least represented in the northern parts of the Old World.

Some members of this pattern, especially in several sections of *Vernonia*, have not only the least differentiated but also the least co-ordinated capitula, in which there are simple and loose involucres and easily discernible individual florets, and which are arranged in the simplest and least regular of compound inflorescences.

6. The wormwood pattern (fig. 147).

The great features here are the small size and considerable aggregation of the capitula, and since these are only comparative terms which do not primarily concern the florets, it will be understood that members of this pattern tend to merge with nearly all the others, except, perhaps, the dandelions, where the capitula are apparently never so numerous. The best practical definition of the wormwood pattern, and it is actually a very satisfactory one, is from the pseudanthial point of view, namely that in this pattern the capitula are of a size, colour and distribution which makes them visually comparable with the true flowers found in many "apetalous" families. For instance in many species of *Artemisia* there is no impression of petaloid flowers, and the inexperienced observer might be excused for taking the plants to belong to such a family as the Chenopodiaceae rather than to the Compositae. At the same time, and this is important, the capitula are not actually in contact with one another, and are sometimes quite widely spaced. The florets may be all discoid or they may include a few inconspicuous ray-florets.

Since the pattern here is based so largely on capitulum size, which is not a point of great taxonomic value, it does not equate with any particular subgroup of the family, and with the reservation made in the last paragraph, representatives are found in most, so that it is, in one form or another, very widely distributed geographically. It is, however, especially typical of *Artemisia* and of the Ambrosinae, the former of which is chiefly in the northern temperate zone, and the latter almost entirely in America.

7. The cudweed pattern (fig. 148).

Here the capitula are so small and their aggregation is so complete (they are normally in contact) that they have no real pseudanthial value, and the inflorescence clusters take the form of spikes or heads which might well belong to such a different family as the Amaranthaceae. In the typical instances of the pattern, indeed, *the capitulum has no real individuality at all and is merely a concealed unit in a capitate or other-shaped mass.* Even more in this pattern may the inexperienced be excused for failing to realize the true affinities of the plants.

It is important to note with regard to both the wormwood and cudweed patterns that the size of the capitula is a general measure of the number of florets they contain, as is true in general terms throughout the Composites. In some at least of the cudweed pattern this reaches its ultimate expression in capitula which contain one floret each only.

The cudweed pattern is fairly widely but rather irregularly

distributed, and is especially characteristic of some of the more arid parts of the northern hemisphere.

Although the seven patterns just described comprise the bulk of the Composites they by no means include all the members of the family. In particular there are many more or less aberrant, in the sense of unaverage, genera and species, and some of these will be noticed in due course. This survey of the patterns has, however, shown that variation in general form within the family is largely due to a small number of characters and especially to differences of life-form and foliage, of number and size of capitula, of the form of the involucre and receptacle, and in diversity among the florets themselves. We cannot deal with all these here, but one or two of them are of special interest and must be considered with some care.

The number of florets in the capitulum
It is difficult to say what the largest number of florets in a capitulum may be in nature, nor is it a matter of great theoretical moment, and it will suffice to remark that in some species of *Helichrysum* it is described as a thousand or more and that the capitula in *Wunderlichia* may be six inches or more across. Much more important is the fact that from this maximum there is, it is safe to say, every appreciable gradation to the opposite extreme where there is only a single floret, but this gradation is uneven in the two respects that certain numbers appear to be more prevalent than others and that the lower the numbers the less variable they are for species or genus. Thus, in larger capitula the numbers may vary considerably from time to time at least (and may be greatly increased by cultivation), but in smaller capitula the number of florets is much more rigidly fixed. This is associated with the fact that the fewer the florets the greater the difference in the outline of the head if one floret is added or subtracted. The smaller the number of florets the less circular the capitulum will become until if the number is only two it can scarcely any longer be described as radially symmetrical. The capitulum with only one floret is best regarded as being without symmetry, though actually the form of the floret will decide whether it is regular or irregular. However this may be there are, from the evolutionary point of view, two particularly interesting

aspects of floret number, namely the prevalence of five or three ray florets and the occurrence of capitula with only one floret.

A good deal of work has been done on the number of ray florets in the capitulum, chiefly from the point of view of phyllotaxis, and Small has given a useful summary of this. The matter can best be introduced here by mentioning Church's contention that, in a complete or many-flowered Composite there is "a rising phyllotaxis", by which is meant that while the foliage leaves are arranged according to a low Fibonacci fraction such as 3/5, the phyllaries may be 8/13, and the disc florets perhaps as much as 21/34. This seems to be merely a rather arithmetical way of saying that these three parts of the plants are, as it were, telescoped in different degrees, or that the two latter are telescoped versions of the first, and this accords with the fact that the phyllaries are usually on the more or less vertical sides of a shallow receptacle, and the florets themselves are on the surface of the receptacle which in many cases at least, and perhaps typically, is more or less completely flat so that they fall into a spiral in one plane. It may also be argued that since there is on flat receptacles no vertical dimension, the larger the number of florets to be accommodated the greater, in a sense, the telescoping must be, and the more complex their phyllotactic arrangement will at least appear to be. At all events it would seem that the peripheral florets in a capitulum correspond to the orthostichies of a leaf phyllotaxis and therefore to the denominator of a Fibonacci fraction. If this is so, then there would appear to be profound organizational reasons why certain numbers of ray-florets such as three or five, should occur in capitula with few florets. In much the same way any major change of numbers, especially at the lower end of the series, would be accompanied by a change in the Fibonacci fraction, in which case a completely "rising phyllotaxis" would result only when the number of florets in the capitulum was large, at least in comparison with the numbers of phyllaries and leaves.

Care must be taken not to read more than is justifiable into the occurrence of particular numbers, but at the same time it would be affectation to deny that two of these numbers, namely 3 and 5, have a peculiar significance because they are, in the guise of trimery and pentamery, the usual numbers for perianth whorls in the Monocotyledons and Dicotyledons respectively, and that consequently any capitulum which has more especially five (for the Composites are Dicotyledons) or even three (the Monocotyledon number) will have a much closer and more convincing resemblance to euanthial flowers than capitula with any other number. That is to say these capitula will be more perfectly pseudanthial and it is merely a statement of fact that, among radiate capitula, those with five or three rays especially reach the highest levels of pseudanthy.

But this raises two other interesting problems, whether there is any real worth or virtue in so high a degree of pseudanthy, and what the evolutionary relation may be between multiflorous and pauciflorous capitula. The first of these is perhaps the more difficult because so much depends upon anthropometric appreciations but the difficulty seems to be this. There is, as we have seen from our survey of the patterns, quite a wide range of pseudanthial form and precision in the Composites. Can we within any reasonable meaning of the word regard the more highly differentiated capitula as being an "improvement" on the others or not? Can we believe that the simpler *Vernonia* condition, for example, is in any material way less surely and adequately functional than that of the thistles; or that either of these is inferior in a biological sense to the daisy or the dandelion; or that one of the two last is biologically superior to the other? The dilemma in this is clear. If we believe any of these things we have to account logically for the continued, and apparently successful, existence of so wide a series of what are *ex hypothesi* inferior types; if we cannot believe any of these things then we must account for the fact that there is so much observable difference between these types. In the latter case it seems to follow that anything in the make-up of a pseudanthium beyond and additional to some minimum requirement is more or less superfluous, or at least gratuitous. That such refinements may, indeed, cause detailed *differences* in function may be confirmed by observation, but to assume that such differences are fundamentally and significantly qualitative is quite another matter.

The number of florets in the capitulum is little used in classification, and the species within large genera often vary greatly among themselves in this respect, so that a laborious calculation would be needed to show the comparative frequency of particular numbers, but it seems safe to say that in most few-flowered capitula the florets are all discoid, and that when there are ray florets it is unusual for them to be as few as five, although where the number is less than ten five seems most common. It is interesting to note that there are two different conditions among such 5-rayed capitula: in one the rays are accompanied by one or more disc florets and in the other there are ligulate (ray) florets only.

The fact that there are five ligulate florets sometimes adds astonishingly to the pseudanthial effect of the capitulum, especially if the rays are broad. Good examples of this are *Chrysogonum virginianum*, where the capitula are not only comparatively large but rather widely spaced: *Sclerocarpus (Gymnopsis) uniserioides*, where the rays are even more ovate and pointed; *Catamixis baccharoides*, where the capitula are very numerous, as they are also in *Nassauvia revoluta*; and *Proustia pyrifolia*. More familiar are the instances in the section Millefolium of the genus *Achillea*, which is notorious for the ease with which its members may be confused, at first sight, with euanthial plants. In *Triptilion capillatum* the rather small capitula are grouped on comparatively leafless stalks, and the rays, which are blue and white, spread so little that the whole ring of them is funnel-shaped. The pseudanthial effect tends also to be heightened when there are no disc florets, as in some of the lettuces (fig. 157), because then the andrecia correspond to the five stamens in many euanthial flowers.

Capitula with fewer than five florets in all have been recorded for at least 75 genera, most of them discoid, while capitula with two florets only occur in more than a dozen. Finally, capitula in which the number of florets is at the absolute minimum, one only (figs. 150), are a good deal more frequent than is sometimes realized and have been mentioned in some or all the species of at least the genera *Acantholepis*, *Ambrosia* (female), *Angianthus*, *Caesulia*, *Corymbium*, *Coulterella*, *Cousinia*, *Echinops*, *Eremanthus*, *Flaveria*, *Franseria* (female), *Gnephosis*, *Gorceixia*, *Haplostephium*, *Hecastocleis*, *Humea*, *Hymenodea*, *Lagascea*, *Lychnophora*, *Myriocephalus*, *Oliganthes*, *Rolandra*, *Spiracantha*, *Stoebe*, *Vanillosmopsis* (fig. 149) and *Vernonia*.

The particularly cogent evolutionary problems raised by these pauciflorous capitula and especially by the uniflorous ones will be discussed in the final chapter, but it may be made plain here that these heads have usually a well-organized involucre in which there is no corresponding lessening of the number of phyllaries. This coupled with the strong indications that capitula are, by and large, telescoped racemose inflorescences, leads almost inevitably to the conclusion that the pauciflorous condition in general and the uniflorous condition in particular, has come about by some process involving a reduction of the number of florets in the capitulum. It should also be noted that such few-flowered capitula, which are naturally relatively inconspicuous, are generally, though not always, borne in considerable numbers and in close aggregation.

Compound capitula

The aggregation of small capitula just mentioned often amounts, as was seen in the case of the cudweeds, to close contiguity, but it is one of the most remarkable features of the Composites that it may go even further than this, and the result is the development of compound capitula, or *pseudanthia of pseudanthia*. It is not easy to make a definitive list of these because three different states are involved. In one, groups of small capitula are not only closely aggregated in space but also actually organized within common involucres; in a second, the aggregation of capitula, though not within recognizable common involucres, nevertheless produces a comparable unification, as is familiar in *Echinops*, the globe thistle; in the third the aggregations of capitula are scarcely circumscribed at all. It is the problem of deciding where the first and third of these begin and end which makes it difficult to say just which Composites have compound capitula, but the list overleaf compiled from several sources, contains virtually all those which seriously merit this description. It will be noted that they belong to nine tribes; that they are unrepresented in the Liguliflorae; and that they exclude such genera as *Evax* and *Ifloga* in which the capitula are no more than unusually close together. Of all these genera *Syncephalantha* is rather different from the rest, and in addition to them there is the genus *Triplocephalium*, in the Inuleae, which is described as having capitula of the third order, or capitula of capitula of capitula. The further evolutionary implications of all these genera will be considered in the final chapter.

Genera of Compositae with Compound Capitula

Vernonieae	Eupatorieae	Inuleae	Cynareae
Vanillosmopsis	*Mexianthus*	*Athroisma*	*Echinops*
(fig. 149)	*Sphaereupatorium*	*Blepharispermum*	(fig. 150)
Gorceixia	Heliantheae	*Sphaeranthus*	*Acantholepis*
Elephantopus	*Lagascea*	*Aetheocephalus*	Mutisieae
Hystrichophora	*Coulterella*	*Pterocaulon*	*Hecastocleis*
Eremanthus	Helenieae	*Monarrhenes*	*Polyachyrus*
Lychnophora	*Syncephalantha*	*Leontopodium*	
Haplostephium	*Flaveria*	(fig. 155)	
Lychnophoropsis	Anthemideae	*Lasiopogon*	
Pithecoseris	*Oedera*	*Chthonocephalus*	
Chronopappus	Arctotideae	*Polycline*	
Soaresia	*Gundelia*	*Decazesia*	
Telmatophila	*Platycarpa*	*Craspedia*	
Rolandra		*Caesulia*	
Spiracantha		*Myriocephalus*	
		Gnephosis	
		Angianthus	
		Cephalipterum	
		Calocephalus	
		Gnaphalodes	

Features of the involucre in the Composites

The involucre, that ring of bracts (phyllaries) which surrounds and contains the florets in the capitulum, is another topic of special evolutionary interest, because it is an involucre, more than any other structure, which is usually the real organizing factor in the pseudanthium (see p. 289 above). Its variation in the family is, therefore, of considerable interest.

An involucre is apparently never totally absent in Composites, or in other words there are no plants lacking an involucre which are otherwise characteristic of the family. It is true that the involucre is sometimes, as in *Wilkesia*, said to be absent but these plants have obvious "involucres" in the visual sense and the statement is based on the proposition that the structure so called here is not strictly homologous with that of other members of the family, being composed rather of the subtending floral bracts of the outermost florets. Similarly in *Eriothrix* the upper foliage leaves are fine and narrow, and pass into the members of the involucre so that it is hard to see any line of demarcation, and it is possible to regard the uppermost either as true phyllaries or as leaves, but in either case the florets are effectively surrounded. If, on the other hand, the involucres of these plants are not anomalous, then all Composites possess one.

The number of phyllaries and the degree to which they are co-ordinated varies a good deal. As might be expected they are notably few in some of the very small capitula, but also in various others too, as *Mikania*, where there may be only four, or *Stevia* where there may be only five, or in some of the Liguliflorae. In a number of genera, among which are *Vernonia, Jaumea, Guizotia, Lasiospermum, Varilla*, and the remarkable genus from the Galapagos Islands, *Scalesia*, there are species in which the phyllaries are not only few but loosely arranged and these may be held to represent the simplest common condition in the family. In this connection *Actinomeris squarrosa* (fig. 154) is noteworthy because the lack of co-ordination is not confined to the involucre, though this is small and so reflexed as to have no enclosing function, and the florets in consequence are loosely held. In addition, the disc florets are long, which intensifies their looseness, there are usually only ten or fewer ray-florets which are reflexed and with considerable gaps between them, and in sum the usual Composite appearance is strongly modified. It may be remarked in passing that the rays are unusually narrow for the size of the capitulum in various other members of the family also, such as species of *Onoseris*. In *Zinnia* and some species of *Inula* the phyliaries are rather more numerous but show little imbrication; in *Hypericophyllum* they are few and broad. They are not uncommonly fused together to some degree, chiefly when their numbers are small as in *Tarchonanthus* (male), *Ambrosia* (male) and *Tagetes*. Sometimes, as in *Cotula turbinata* where the upper part of the peduncle gradually dilates it is hard to say how far there is fusion: sometimes the phyllaries though numerous, little overlap, as in *Bigelowia graveolens*. In the striking *Stifftia (Augusta) chrysantha* the involucre is very abbreviated compared with the long florets and pappus and this contributes to an unusual appearance.

At the other end of the numerical scale *Cnicothamnus* is reported to have more than 200 phyllaries, and in *Tricholepis* they are not only very many but have spreading hair points, but something much more modest is the norm. In these more average types variation consists chiefly in the arrangement of the phyllaries, and particularly how much imbricated and alike they are. In *Helianthus*

all are alike in more than one row; in others, especially Coreopsi-
dinae, and in such familiar species as *Senecio vulgaris* there are two
rows the outer much smaller than the inner; and in the equally
familiar *Taraxacum officinale* the outer phyllaries are, in the late
bud stages, spread out so as to form a green perianth-like surround
to the rest of the unfolding capitulum. Generally speaking,

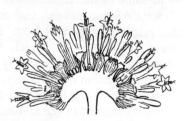

FIG. 149. *Vanillosmopsis erythropappa:* section
of compound capitulum, considerably
enlarged, after Baillon

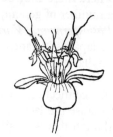

FIG. 151. *Pinillosia
tetranthoides:* capi-
tulum, × 5, after
Baillon

FIG. 150. *Echinops sphaeroce-
phalus:* one-flowered capitulum (left), × 2; single floret
(right), × 2

however, the more numerous the phyllaries the more regularly
imbricated they are, as in the thistles and the even more notable
Barnardesia rosea. In such cases the phyllaries commonly all arise
at much the same level so that the inner ones are progressively
longer. Other good examples of this kind of involucre are *Haplo-
phyllum, Lychnophora, Liatris, Catananche, Phaenocoma prolifera*
and *Achyrocline satureioides.*

Another interesting variation is the degree to which the involucre
is narrowed at the neck, as in many thistles and, more markedly,
Centaurea salmantica, Silybum marianum and the globe artichoke
(*Cynara scolymus*). These are the antithesis of such as *Helenium*
where the involucre spreads almost flat, and where this is often

accompanied by an increased convexity of the receptacle, the one substituting as it were for the other.

It is in the more deeply imbricated urceolate involucres that the apical specialization of the phyllaries is best seen. In detail this is various but by and large it takes the form either of pinnate feathering, as in *Centaurea* spp., or of spines as in the thistles and other *Centaurea* spp., or more rarely of hooks, as in *Arctium*. Some urceolate involucres, especially if they are robust, serve to protect the ripening fruits, and as in the last-mentioned may be disseminated whole. Finally, there are several quite extravagant specializations, also associated with protection and dissemination. The best known is perhaps the female capitulum which develops into the hooked "fruits" of *Xanthium*, the cocklebur, but *Acanthocephalus*, a member of the Liguliflorae in which the true fruits also contribute to the general effect, is even more noteworthy. It may be added that protection of the fruits is also afforded by various quite different and simpler involucres as in *Monactis*, *Rumfordia*, *Lagophylla*, *Layia*, *Jaegeria* and *Zacyntha verrucosa*.

Colour, other than green, is best known in the involucre from the scarious phyllaries of *Helichrysum*, *Carlina* and others of the everlasting-flower pattern, which not only function like ray florets but which are also protective to some degree through their hygroscopic movements. Usually it is the inner or adaxial surfaces of the phyllaries that are most brightly coloured but sometimes the reverse is true, or they may be coloured on both sides, as in *Rhodanthe manglesii*. In two species of *Cremanthodium* (fig. 156), *C. wardii* and *C. campanulatum*, the capitula are solitary, cernuous and suburceolate, contain only disc florets, and have little more than a simple row of phyllaries, but these are softer in texture than usual and pale to darker purple in colour. The whole is strongly reminiscent of certain quite other pseudanthia such as those of some species of *Darwinia* (see p. 289 and fig. 137). In a few instances, notably *Vernonia corymbosa*, the innermost phyllaries are in posture, texture and general appearance much more like ordinary corolla ligules.

The involucre tends particularly to have specialized form in Composites with very pauciflorous capitula, though much depends on whether these are simple or compound. In the former the one-flowered capitulum is of particular significance because this contains only one fruit and not unnaturally often behaves as a single complete disseminule, an end to which the involucre may contribute. In *Ambrosia* and *Franseria* it is partly spiny for example, and in *Iva* partly membranous. It is, however, in the

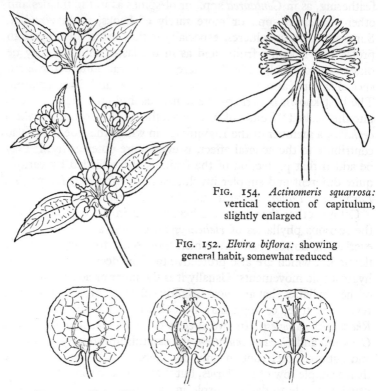

FIG. 154. *Actinomeris squarrosa:* vertical section of capitulum, slightly enlarged

FIG. 152. *Elvira biflora:* showing general habit, somewhat reduced

FIG. 153. *Elvira biflora:* showing three aspects of a capitulum, × 3

group of the Millerinae, which are pauciflorous though not uniflorous, that the most remarkable states of the involucre are to be seen. All the dozen or so genera are worth careful examination from one or other point of view here, but space permits reference to only three of them. In *Pinillosia* (fig. 151) the capitulum has only four flowers, two hermaphrodite and two female, and the involucre has six phyllaries of which four, much larger than the other two, are spread cruciately, the florets projecting far beyond them. In *Milleria* the small capitula are often zygomorphic because they may contain among their few florets only one which is ligulate, and the very simple involucre, which has one member much larger than the rest, becomes fleshy and envelops the fruit as the latter ripens. In *Elvira* (figs. 152, 153), which may claim to be one of the strangest of all Composites, the capitula have only a very few (often two) simple male or female florets but the involucre is of a few erect and accrescent net-veined "spuriously samaroid" phyllaries, the two larger of which are compressed flatly against one another and entirely conceal the florets within. As a result the capitula, which are rather numerous and irregularly grouped, appear much like the flowers or young fruits of such a plant as *Atriplex*, and all trace of the Composite facies is lost. A rather similar condition is to be seen in *Staurochlamys* and in *Dicoria calliptera*.

There are three particular points about the involucre in Composites with compound capitula. One is that in some of these plants e.g. *Vanillosmopsis* (fig. 149) and *Sphaeranthus* there is a true compound involucre as highly organized as that in many simple capitula. Another is that even when the compound capitula are best defined, as in *Echinops* (fig. 150), the involucres of the individual capitula may be quite normal for their kind. The third is that in certain genera, often classified together as the Agianthinae (all of which are under Inuleae in the above list) the co-ordination and suppression of individuality in the capitula may be such as to result in a compound inflorescence very uncharacteristic of the Composites in general, and in which also the involucres lose much of their normal nature. *Angianthus* itself is one of the best examples of this, for in some species, e.g. *A. myosuroides*, the compound inflorescence can best be described as an elongated cone of bracts each subtending a small true capitulum with its appropriate involucre, which last is itself much modified.

The case of *Leontopodium* (fig. 155), which is also in the list above, deserves mention because here the bracts of what in fact constitute the common involucre are radiate and flannelly in a way which almost merits the term perianth-like, and this is particularly interesting because were it not for the peculiarities of these bracts, the genus could scarcely be included in the list, since the constituent true capitula are quite clearly defined.

Finally, with regard to the involucre, and clearly allied to the state of *Leontopodium*, are the cases in which capitula are subtended by leafy bracts which can often be regarded as a second and outer involucre, as is seen in *Bidens cernua* and *B. tripartita* and in *Vernonia pedunculata*, *Centratherum intermedium*, *Spiracantha cornifolia*, *Pertya scandens*, *Pallenis spinosa* and *Distegia*. These bracts may be very small, as in *Orochaenactis thysanocarpha* and

FIG. 155. *Leontopodium alpinum:* somewhat reduced

FIG. 156. *Cremanthodium campanulatum:* × ½

Otopappus verbesinoides, or specialized, as in the glandular ones of *Siegesbeckia droseroides*. A very unusual condition is that of *Caesulia axillaris* in which the *compound* capitula are sunk deep in the leaf axils and surrounded by coloured bracts.

Features of the pappus and of the receptacle bracts in the Composites
Space allows reference to only one major aspect of the pappus, namely its variable development in the family. The number of

Composites which are without a pappus is hard to compute, partly because it is not easy to say at what stage functional value begins; partly because the pappus may be caducous, so that even if it is sometimes conspicuous its functional value cannot be assumed; and partly because related individuals are not always alike in this respect. In certain genera a pappus of some sort is described in some species but not in others; it may also be absent from some varieties of plants which normally have it; there are plants in which it varies still more arbitrarily among individuals; and there are still others in which some florets of the capitulum have a pappus while the rest have none. Neighbouring genera may vary greatly in respect of the pappus (see below) but this may be due to faulty taxonomic attribution.

On a wider scale every tribe of the family contains some plants without a pappus, for even in thistles, where the "down" is so familiar, there are species of *Centaurea* which have none, but there is no tribe in which all the plants lack a pappus. In summary, the Anthemideae and Calenduleae are nearly all without pappus, and this condition is common also in Heliantheae, Helenieae, Inuleae and Astereae. The tribes, however, vary in size, and taking this into account it would seem that about one third of all the genera, and probably not much more, are without a pappus recognizable as such. But these genera, again, vary in size, and include only two which are really large, *Achillea* and *Artemisia*. It thus seems clear that the great majority of all Composite species have a reasonably well-developed, and presumably functional, pappus, this being usually of the "thistle-down" type.

Two not uncommon garden plants, *Inula helenium* (elecampane) and *Buphthalmum speciosum* (ox-eye), show a remarkable combination of resemblance and contrast involving both pappus and floral bracts. They are tall, coarse, leafy perennials with large yellow-rayed capitula, with few little-imbricated and rather leafy phyllaries, and taxonomy assigns them, in close proximity, to the tribe Inuleae. But here the resemblance ceases for within the capitula not only are the ligules rather different in posture and colour, but while in the *Inula* the achenes are pale-coloured and have a well-developed and persistent hairy pappus, in the *Buphthalmum* they are dark and virtually without a pappus at all. Furthermore, while the *Inula* is without any floral bracts on the receptacle, these are extremely marked and persistent in the *Buphthalmum*, so much so, indeed, that the withered capitula, after the achenes have fallen, look almost like small flat teasels. Thus, of the two species, which in general silhouette and appearance are so much alike, one has a well-developed pappus but no floral bracts, and the other has well-developed floral bracts but no pappus.

In *Echinacea* (*Rudbeckia*) *purpurea* the striking orange-red colour in the disc, which is in strong contrast to the pink ray florets, is due to the brightly coloured rigid tips of the floral bracts, which project well beyond the small and inconspicuous green disc florets.

Pendent and horizontal capitula

One of the points about the Composites most unexpected to those not familiar with it is the occurrence here and there in the family of species with nodding or even fully pendent capitula, these two adjectives in general denoting two degrees of the same tendency. The former are illustrated by one or two familiar European plants such as *Bidens cernua* and *Carduus nutans*, in both of which the capitula are comparatively large and multiflorous, and in the African *Erythrocephalum nutans*. The latter may be exemplified by the two species of the Asiatic genus *Cremanthodium* already mentioned; by species of the northern temperate genus *Prenanthes*; and by the South American genus *Mutisia*.* Of *Prenanthes* a brief account has been given elsewhere (*Journal of Botany*, 1931) but it is of such interest that the leading facts deserve repetition. The account refers particularly to one species, *Prenanthes purpurea*, though there is at least one other, *P. violaefolia* from Nepal, which is very similar.

P. purpurea (fig. 157), which is not uncommon in Europe and especially in the Alps, is a tall perennial herb of the lettuce kind, with pale purple corollas and with the capitula arranged in rather lax axillary panicles of a dozen or more well-spaced capitula. There are three remarkable features about these capitula. First, their styles point towards the ground and they may, therefore, be said to be fully pendent or inverted, though this must be qualified by the statement that although the capitula hang down they seem to do so more because of the direction of growth of their axes than

---

* See also *Lachanodes prenanthiflora* from St. Helena.

because of any lack of strength in these. Second, there are only five florets in each capitulum, and since *Prenanthes* belongs to the Liguliflorae there are no disc florets. The ligules are usually a little reflexed at anthesis. Third, the anthers, and later the styles, are exceptionally prominent, the former being completely exposed beyond the involucre. The effect of the inversion of the capitula, the presence of only five, all ligulate, florets, and the protrusion of the anthers and styles, is to give a floral unit which, in its manner of function, is quite foreign to that of the ordinary Composite, and closely parallel with many euanthia such, for example, as the

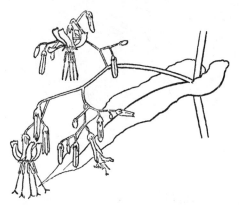

FIG. 157. *Prenanthes purpurea:* portion of the compound inflorescence, somewhat reduced

flowers of *Fuchsia* or some species of *Silene*. Moreover, the capitula are pollinated in somewhat the same way, by medium-sized species of humble-bees which maintain themselves in position during the process by clutching the protruding parts.

The genus *Mutisia* consists chiefly of climbing plants with tendrils and large solitary terminal capitula. In some species, but not all, and especially in those where the peduncles are long, the capitula are fully pendent, and on this account, and because of their size, present a remarkable appearance (fig. 158). In these capitula the anthers project wholly beyond the level of the rays, and the styles grow out still further, so that here again the whole functions as a single large pendent flower with projecting stamens and styles. In this case, however, the capitulum contains a considerable number of disc florets, so that the number of stamens involved is also larger, and the parallel is closer, perhaps, with some polystemonous flowers, such as members of the Myrtaceae, Malvaceae or Bombacaceae. These fully pendent kinds of *Mutisia* and *Prenanthes* may be regarded as the antithesis, at least as regards pollination method, of the usual larger complete Composite

FIG. 158. *Mutisia grandiflora:* capitulum, × ½, after *Pflanzenfam.*

such as many of the Heliantheae, in which there are erect capitula with large numbers of disc florets so nicely and closely packed that they provide an even surface on which pollination is accomplished by the crawling of insects. Yet the organizational ingredients of these two sorts of plants are homologous and differ only in detail.

In another remarkable Composite, the Asiatic species *Ligularia przewalskii*, which is sometimes grown in gardens, there is a state of affairs which, although not quite the same, is quite as remarkable as that just described. This plant is a large perennial herb reaching five or six feet, with large palmatisect basal leaves (fig. 159). Its capitula (fig. 160), which though not aggregated, are

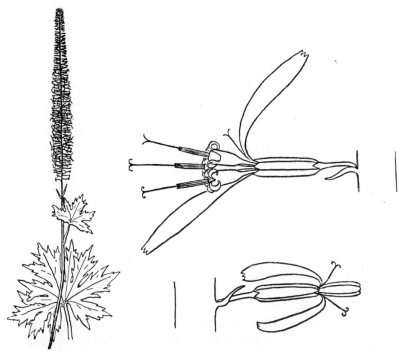

FIG. 159. *Ligularia przewalskii:* habit, × 1/10

FIG. 160. *Ligularia przewalskii:* young and (above) older capitula, × 2½

very numerous, are arranged in long erect terminal multiradiate *racemes*, each of which is a direct continuation of an almost leafless axis rising from the base of the plant and may bear as many as 300 or 400 capitula. The great feature of these is that, although shortly pedunulate, they hold themselves more or less strictly horizontal throughout their life, and thus confer on the raceme as a whole a most unusual appearance, and one for which it is hard to think of a parallel. Moreover, each capitulum has a long cylindric few-phyllaried involucre and contains five florets only, three being hermaphrodite disc florets and two female ray florets. The corollas of the disc florets are narrowed below and are campanulate above and the lobes are re-rolled, and all the campanulate part as well as some of the tube is exposed beyond the involucre, while the anthers are in turn fully exposed beyond the corolla mouth. A curious point is that the campanulate part of the disc florets is of a different consistency from the rest and changes colour to brown long before the ligules of the ray florets begin to fade. The position of these two ray florets varies from capitulum to capitulum but they are usually either on the functional "upper" side, where they form a conspicuous yellow V, like a pair of wings, or they are on the "upper" and "under" sides of the capitulum respectively, so that the whole is bilaterally symmetrical along a vertical axis. In the latter circumstance the two ligules generally show the remarkable condition that the upper one is reflexed or bent back towards the axis of the raceme, while the lower one projects rather straight forward. The effect is to make the capitulum appear bilabiate, or at least to have that degree of oblique zygomorphy commonly seen in ordinary flowers with a horizontal posture, and this parallel in the case of a Composite is of great interest. The florets are visited by humble bees, and doubtless others of the larger Hymenoptera.

Cosmos and Anemone

This chapter may well conclude with a comparison which serves to epitomize many of the evolutionary problems in the Composites.

The complete Composite capitulum being interpreted as a pseudanthium, that is to say as functionally and visually a single flower, it may be described as a reproductive structure of a design which determines that, if it fulfils its function efficiently, it will achieve the biological end of producing one-seeded indehiscent fruits (disseminules) equal in number either to the totality of its florets or at least of some considerable proportion of them. A moment's consideration will suffice to recall that there is nothing peculiar to the Composites in this and that there is in fact exactly the same culmination in certain other plants. Not only so but this is notably true of a family which we are commonly invited to think of as the structural and taxonomic antithesis of the Composites namely the Ranunculaceae, for almost any single member of the tribe Anemoneae there has single flowers of which almost exactly the same can be said. A particular comparison of this kind is that between *Cosmos* (*Cosmea*) *bipinnatus* and *Anemone japonica*, which are often to be seen growing together in gardens. The reproductive entities of these two are alike in size, posture, shape, colour and general proportions (figs. 161, 162) and each has, as its

Fig. 161. *Anemone japonica:* plan of flower, about half natural size

Fig. 162. *Cosmos bipinnatus:* plan of capitulum, about half natural size

biological end, the production of a number of one-seeded inde-hiscent disseminules without any special "dispersal mechanism" so that it is most difficult to detect any biological disparity between them. Indeed, the advantage, if one may use that word, even seems to lie with the *Anemone* because it produces on the average more embryos *per* flower than the *Cosmos* does *per* capitulum. It is, therefore, hard to see what possible "benefit" the Composite form, and the presumed immense amount of evolutionary change that has gone to its elaboration, confers upon its possessor. Certainly there are true fruits in the former and false fruits in the latter but such details as these can scarcely be held to weaken the comparison.

NOTE ON LITERATURE

For further information of the more formal sort about the Composites the general taxonomic works cited at the end of Chapter 4 should be consulted.

Two other important sources about the family are:

Bentham, G.: "Notes on the classification, history, and geographical distribution of Compositae." *Journ. Linn. Soc. (Bot.)*, **13**, 1873.

Small, J.: "The origin and development of the Compositae." *New Phytologist*, **16–18**, 1917–1919. Reprinted 1919.

The latter is specially valuable for its extensive bibliographies. Regarding its taxonomic and phylogenetic conclusions it is useful to consult the review by Spencer Moore in *Journal of Botany*, **58**, 1920.

The paper alluded to on p. 318 is :

Good, R.: " Some evolutionary problems presented by certain members of the Compositae." *Journal of Botany*, **69**, 1931.

CHAPTER ELEVEN

REPETITION AND SUPERFICIAL RESEMBLANCE

In the course of the preceding chapters comment has several times been made on the nature of the differences between individuals in the organic world. It has been said, for instance, that the larger these differences are the less often they are manifested, and also that most individuals are extremely like many other individuals, while, further, the opinion has been expressed that the multiplicity of the Flowering Plants, which is of course the upshot of such individual differences, is due to the occurrence of many characters in many values and the absence of any general system of correlation between them. Each of these comments has its own particular usefulness but they are all partial expressions of the fundamental biological fact that individuals are like or unlike one another to almost every degree, both quantitative and qualitative. The subject of this disparity is, therefore, one of great complexity.

So complicated is it, indeed, that it has inevitably invited that process of simplification as an aid to human analysis which was discussed at some length in Chapter 4. In this particular case simplification takes several forms but comment may be limited to the two of them which are of special interest here. The first of these is the concentration of attention upon *difference* rather than upon similarity, which has come about partly because disparity is much more easily assessed and described than likeness, and partly because the purpose of taxonomic classification is essentially discriminatory. The second is the recognition, with some sharpness, of two sorts of resemblances, those which can be ascribed without undue difficulty to genealogical consanguinity, and which are therefore *intrinsic*, and those which cannot be so ascribed and which are, therefore, to be regarded as *extrinsic* in origin. The latter are usually called *superficial*, and because of the widespread and long-existing pre-occupation with phylogeny in evolutionary studies such superficial resemblances have commonly been regarded as of relatively minor interest. In consequence they have been much neglected and it is the purpose of this chapter to consider afresh some of the many relevant points about them.

We may well begin by enquiring how far this simplification into intrinsic and extrinsic (superficial) resemblance is a real and useful dichotomy. It is true that resemblances of certain kinds (that is to say which involve certain characters in particular) have, for a variety of reasons, gradually come to be accepted as more reliable indicators of kinship than others, but it is to be remembered that this belief is by no means as firmly founded as may at first appear. Thus, it is difficult to mention any *single* character which can confidently be accepted in all circumstances as indicating consanguinity among its possessors, and therefore as wholly intrinsic rather than extrinsic, and indeed, it may be argued that the whole fabric of modern plant classification rests on an implicit acceptance of this view. On the other hand it is a logical supposition that the greater the totality of resemblance the greater the likelihood that it it is due to true kinship, and it is, in fact, the amount or quantity of resemblance, rather than its quality, which is commonly made the main criterion of relationship, though the actual nature (or quality) of the characters involved affords an important secondary criterion. In short, if two plants resemble one another in a *series* of more *particular* features, such as are not of common occurrence elsewhere, then they may be assumed to be related in blood: if, however, they resemble one another in only one or two characters, these are not, however particular they may be, generally considered necessarily to indicate close kinship. It would appear then that we cannot really recognize two sharply contrasting kinds of characters, extrinsic and intrinsic, and hence that the use of resemblance in estimating relationship must be largely empirical. Both kinds of resemblance are due, as is inevitable, to the same general phenomenon of nature, which is the occurrence of the same character or combination of characters in a variety of individuals, and hence individuals may be like one another to almost any degree. They may be alike in one character only or in any number of characters, the greater the number the nearer being the approach to identity, and the nearer the approach to identity the greater the probability of true relationship. When we speak of superficial resemblance we therefore mean, in effect, no more than resemblances which are, in the circumstances of the case, insufficiently considerable to allow of the attribution to them of genealogical significance. They may also be described as resemblances which, however interesting they may be in themselves, are less important in this respect than are the differences which accompany them. It is with such definitions as these that we begin to appreciate the real point of evolutionary interest in superficial resemblances, which is why, if they are as extrinsic as has been suggested, they should occur at all.

From these definitions, too, it is to be seen that superficial resemblance is likely to express itself in a very wide range of values, because a likeness in any readily visible character, however insignificant, must inevitably confer some degree of visual similarity, and that in consequence this kind of likeness may be especially difficult to analyse. This is true, but there is to hand an empirical simplification which can be of considerable practical service. Inseparable from the ordinary idea of superficial resemblance is a degree of likeness which results, at least to the eye, in some measure of *generalized* similarity, but there are many degrees of extrinsic resemblance which cannot claim to amount to this. This is especially the case when only one feature, or it may be a very small number of features, is concerned, and it is helpful, therefore, in these pages at any rate, to confine the term superficial resemblance to the former, more generalized, phenomenon, and to treat the latter rather separately. This latter, which in its most typical expression takes the form of resemblance in a single character only, is, for that very reason, of special interest in relation to the science of genetics, but it has another aspect which concerns us even more closely here. The occurrence of the same relatively unusual character in two or more widely differing lineages or natural groups, in any or all of which it may itself be uncommon, gives an impression, which is hard to dispel, that the character in question has appeared *de novo* more than once in the course of change with time. This idea of *repetition* is of course at the bottom of all superficial resemblance, which could not exist without it, but in the more generalized phenomenon the combined result of a number of similarities tends to minimize the separate visual effect of any one of them, so that the impression of repetition is not so striking. It seems justifiable, therefore, to give the name *repetition* here to the occurrence of similarity in one or very few characters only and to deal with it first and separately.

Repetition

Just what is meant in this chapter by repetition and its relation to evolutionary problems can, in view of what has already been said, most conveniently be illustrated concisely by describing four examples of it, each of a rather different sort.

1. *Flower colour*

The general distribution of flower colour in the Flowering Plants exemplifies the kind of repetition which is of such widespread occurrence that its essential nature is apt to be overlooked. Nearly every group of flowering plants has its own typical range of flower colour, to which the majority, at least, of its members conform, but it is, nevertheless, seldom if ever possible to recognize any but some of the smaller groups unfailingly on flower colour alone. Nor are there many instances of unique flower colour, even if simple forms of patterning are included, and the unusual state of affairs described earlier in the Asclepiads and Orchids is for this reason of special interest. In fact repetition in flower colour is a commonplace and tends only to strike the eye when the tints concerned are relatively uncommon, as, for instance, in some of the exotic plants with brick-red flowers such as *Dimorphotheca*, *Ursinia* and *Ixora*, or in the more striking cases where there is a conspicuous combination of red and yellow, as in:

> *Manettia ignita* (Rubiaceae)
> *Siphocampylus manettiaeflorus* (Lobeliaceae)
> *Ceratostema speciosum* (Vacciniaceae)
> *Desfontainia spinosa* (Desfontainiaceae)
> *Scutellaria costaricensis* (Labiatae)
> *Epidendrum radians* (Orchidaceae)

An interesting point about flower colour is that it is so directly interpretable in chemical terms. Nearly all flower pigments are either cell-sap pigments (usually seen as ivories, pale yellows, pinks, reds, blues and purples) or plastid pigments (mostly seen as strong yellows and oranges), and some of the less common effects result from the simultaneous occurrence of both these kinds. More generally, however, a group is characterized by one or the other, though a few, of which the Dahlias are a familiar example, include both. There is also a close connection between colour and the acidity or alkalinity of the cell-sap, and changes in flower colour such as occur in many Boraginaceae, or in the spots of horse-chestnut flowers, are generally associated with change in hydrogen-ion concentration. However, the fundamental reason for the frequency of repetition in flower colour is of course that the total of observably different colours and combinations is much smaller than the number of different floral types or groups, so that many of the latter must be alike in terms of the former. But if this were the whole of the story there might, nevertheless, be perfect taxonomic segregation in flower colour, and the fact that there is not, and that to a remarkable extent, almost any flower colour may be found, somewhere or other, in almost every large group, is one of the most easily appreciated illustrations of what is meant here by the repetition of characters.

2. Succulent or fleshy fruits

This example may be regarded as a morphological version of the first and shows how there is, in the Flowering Plants, a similarly unsegregated distribution of many structural values. Repetition is, indeed, particularly well seen in fruit features, and a second instance of it will be discussed next, but for the moment concern is with the occurrence of fleshy fruit sorts here and there in groups of plants where they are not the rule. This has four noteworthy expressions. First, there is the condition in which only one or two members of a group have fleshy fruits and are, therefore, anomalous among their congeners. *Cucubalus* in Caryophyllaceae and *Actaea* in Ranunculaceae are familiar instances of this, and among others are *Elaeophorbia* (Euphorbiaceae); *Malpighia* (Malpighiaceae); *Palisota* (Commelinaceae); and *Hydrocera* (Balsaminaceae). Less marked examples are *Leucocarpus* and a few other genera in Scrophulariaceae, and *Canarina* and one or two more in Campanulaceae. In *Begonia* there is a similar but intrageneric difference a few species only having fleshy fruits. It is interesting to mention the contrasting state of affairs which is also sometimes to be seen, as in the family Taccaceae, where the monotypic genus *Schizocapsa* has, unlike the larger genus *Tacca*, a capsular fruit.

Second, there are the cases where there is a distinct and well-defined minority of fleshy-fruited plants in groups where other sorts are more largely representative, as in Aizoaceae, Amaranthaceae, Convolvulaceae, Goodeniaceae, Lobeliaceae, Onagraceae, Resedaceae and Violaceae.

Third, there are groups wherein the two conditions are more nearly balanced, the best-known of these perhaps being the Myrtaceae, where one subfamily has fleshy fruits and the other dry. In Bromeliaceae the true fruits from superior ovaries are generally capsular and the false fruits from inferior ovaries are fleshy. Other families in which both sorts of fruit are well shown include Ericaceae, Melastomataceae, Passifloraceae, Rubiaceae and Tropaeolaceae. Pandanaceae are notable for having what are in effect two kinds of fleshy fruits.

Fourth, and from the evolutionary point of view most interesting, are the groups in which, although one particular kind of dry fruit is extremely characteristic, there are, nevertheless, to be seen occasional departures from this towards a condition of succulence. Two remarkable cases here are those of *Melocanna* and one or two other genera in the Gramineae, and *Andira* (also with one or two others) in the Papilionaceae, but much the same sort of thing is seen in the Asclepiadaceae (see Chapter 7), Labiatae (Prasioideae), Malvaceae (*Malvaviscus*) and Orchidaceae (*Vanilla*), and there is even something of the kind in Compositae and Cyperaceae.

3. *Winged fruits and seeds*

This is a more complex example than either of the two preceding. As regards winged *fruits* there are basically two plans of construction to be seen. In one, which may be called the "calyx" type, the expanded surfaces are the persistent and enlarged sepals or corresponding structures. In the other, which may be called the "flanged" type the wings are outgrowths of one sort or another from the surface of the fruit.

The calyx type of winged fruit is seen in most striking form in the family Dipterocarpaceae and in *Lophira* in the Ochnaceae, though similar but less robust instances are found in *Triplaris* (Polygonaceae), *Engelhardtia* (Juglandaceae), *Ceratopetalum* (Cunoniaceae), *Calycopteris* (Combretaceae), *Porana* (Convolvulaceae) and *Kissenia* (Loasaceae). These, in turn, are approached by *Gyrocarpus* (Hernandiaceae), *Homalium* (Flacourtiaceae), *Parishia* and *Swintonia* (Anacardiaceae), *Hiptage* and *Thryallis* (Malpighiaceae), *Patagonula* (Boraginaceae) and the rather similar *Alberta* and *Gaillonia* (Rubiaceae). *Arctotis*, *Berkheyopsis* and *Ursinia* among Compositae are also essentially the same. Modifications of the plan are seen in *Neuropeltis* (Convolvulaceae) where the calyx forms a single round wing, in *Hildebrandtia* of the same family where it forms two such wings, and, rather differently, where the sepals are rounder and more spreading, as in *Piptoptera* and others in Chenopodiaceae and in *Atraphaxis* (Polygonaceae). In *Erisma* (Vochysiaceae) only one sepal is much enlarged.

Expressions of the flanged plan in winged fruits are more numerous both in kind and occurrence. First may be mentioned those in which there are three or more, very often four, flanges or narrow wings running the length of the fruit such as is seen, associated with various dehiscence, in:

Asclepiadaceae	*Dregea abyssinica, Exolobus patens*
Begoniaceae	*Begonia*
Bombacaceae	*Cavanillesia*
Burmanniaceae	*Burmannia*
Burseraceae	*Triomma*
Caesalpiniaceae	*Cassia*
Celastraceae	*Euonymus, Wimmeria*
Combretaceae	*Combretum, Terminalia, Quisqualis*
Compositae	*Tripteris sinuata*
Cunoniaceae	*Gillbeea*
Cyrillaceae	*Cliftonia*
Dioscoreaceae	*Dioscorea* (see p. 334)
Euphorbiaceae	*Manihot*
Flacourtiaceae	*Grandidiera, Poggea*
Gentianaceae	*Exacum*
Haloragaceae	*Loudonia*
Hernandiaceae	*Illigera*
Lecythidaceae	*Petersia*
Malpighiaceae	*Jubelina, Mascagnia et al.*
Melastomataceae	*Bertolonia*
Meliaceae	*Aitonia*
Mimosaceae	*Tetrapleura*
Nyctaginaceae	*Abronia, Phaeoptilum, Selinocarpus*

Papilionaceae	*Piscidia, Psophocarpus, Sesbania, Tetragonolo-bus et al.*
Rhamnaceae	*Gouania, Pleuranthodes*
Rubiaceae	*Kadua*
Sapindaceae	*Paullinia, Urvillea*
Stackhousiaceae	*Stackhousia*
Stenomeridaceae	*Stenomeris*
Sterculiaceae	*Kleinhovia*
Tiliaceae	*Columbia*
Umbelliferae	*Angelica, Asteriscium, Laretia, Laserpitium, Prangos, Pyramidoptera, Thapsia et al.*
Zygophyllaceae	*Bulnesia, Tribulus*

In *Centrostegia thurberi* (Polygonaceae) there is a similar winged involucre.

A corresponding state but with only two flanges occurs, among others in:

Combretaceae	*Terminalia*
Compositae	*Coreopsis, Matricaria, Verbesina, Zexmenia*
Mimosaceae	*Gagnebina*
Simaroubaceae	*Soulamea*
Styracaceae	*Halesia*

but such pass imperceptibly into the more truly two-winged states seen in:

Betulaceae	*Betula*
Boraginaceae	*Heliotropium pterocarpum*
Chenopodiaceae	*Grayia*
Combretaceae	*Anogeissus, Pteleopsis*
Rutaceae	*Helietta*
Urticaceae	*Memorialis*

A notable variant of both the above is when the flanges or wings occur only *below* the carpels as in:

Araliaceae	*Myodocarpus*
Polygonaceae	*Brunnichia*
Sapindaceae	*Serjania exarata, Toulicia*

In the other direction the two-flanged state passes into that of the well-developed, but usually incomplete, border right round the fruit, as in:

Aceraceae	*Dipteronia*
Caesalpiniaceae	*Haematoxylon*
Cruciferae	*Aethionema, Biscutella, Peltaria, Psychine, Thlaspi et al.*
Eucommiaceae	*Eucommia*
Euphorbiaceae	*Hymenocardia*
Juglandaceae	*Pterocarya*
Malpighiaceae	*Hiraea*
Papilionaceae	*Dalbergia, Pterocarpus*
Polygalaceae	*Monnina*
Rutaceae	*Ptelea*
Simaroubaceae	*Ailanthus, Desbordesia*
Ulmaceae	*Ulmus*

Such peripheral winging is seen even in nutlets, as in *Perilomia* (Labiatae) and *Rindera* (Boraginaceae). In *Paliurus* (Rhamnaceae) there is the rare condition that the wing is horizontal: in *Dobinea* (Anacardiaceae) the wing is a bract: in *Artedia squamosa* (Umbelliferae) and *Brachycome calocarpa* (Compositae) the wings are repeatedly and deeply lobed.

Finally, there is the true "key" shape, where there is only a single wing at one side or one end of the fruit or mericarp, and among examples of this are:

Aceraceae	*Acer*
Anacardiaceae	*Loxopterygium, Schinopsis*
Celastraceae	*Plenckia*
Dioscoreaceae	*Rajania*
Magnoliaceae	*Liriodendron*
Malpighiaceae	*Banisteria*
Malvaceae	*Hoheria*
Oleaceae	*Fraxinus*
Papilionaceae	*Centrolobium, Ferreirea, Machaerium*
Phytolaccaceae	*Gallesia, Seguieria*
Polygalaceae	*Securidaca*
Rhamnaceae	*Ventilago*
Sapindaceae	*Diatenopteryx, Thouinia*
Sterculiaceae	*Tarrietia*

In the Papilionaceous genera *Toluifera* and *Platypodium* the wing is basal.

Turning now to winged *seeds*, there is, for obvious reasons, nothing corresponding to the "calyx" type, nor are there basal wings, but there are three conditions which correspond closely to some of those just described.

Seeds with two opposite wings in the same plane are exemplified by *Stereospermum* (Bignoniaceae) and by *Dioscorea* (Dioscoreaceae).

Seeds with a more or less complete peripheral wing or broad border are to be found, amongst others, in:

Annonaceae	*Oxymitra*
Apocynaceae	*Allamanda, Aspidosperma, Plumeria*
Bignoniaceae	*Anemopaegma, Campsis, Delostoma et al.*
Cruciferae	*Farsetia, Parrya, Vesicaria et al.*
Cucurbitaceae	*Zanonia*
Dioncophyllaceae	*Dioncophyllum*
Dioscoreaceae	*Dioscorea*
Fouquieriaceae	*Fouquieria*
Gentianaceae	*Voyria*
Goodeniaceae	*Velleia*
Hypericaceae	*Hypericum*
Loganiaceae	*Antonia*
Meliaceae	*Khaya*
Proteaceae	*Hakea*
Rubiaceae	*Bouvardia, Cinchona, Hindsia et al.*
Rutaceae	*Dictyoloma*
Sapindaceae	*Magonia*
Sarraceniaceae	*Heliamphora*
Saxifragaceae	*Montinia*
Violaceae	*Anchietea*

In *Nepenthes* (Nepenthaceae) and in *Leiphaimos* (Gentianaceae) the winging is exceedingly narrow and elongated, as it is also in *Ourouparia* (Rubiaceae): while *Moringa* (Moringaceae) has seeds with three wings or flanges. In *Paulownia* (Scrophulariaceae) the seed has many wings.

Seeds of the "key" form, some of which are superficially very like fruits, are to be seen, for example, in:

Bombacaceae	*Bernoullia*
Celastraceae	*Kokoona*
Cunoniaceae	*Belangera*
Dioscoreaceae	*Dioscorea*
Hamamelidaceae	*Bucklandia*
Hippocrateaceae	*Hippocratea*
Meliaceae	*Cedrela, Elutheria et al.*

Onagraceae	*Hauya*
Proteaceae	*Banksia, Embothrium*
Rubiaceae	*Sickingia*
Rutaceae	*Terminthodia*
Theaceae	*Gordonia*
Tiliaceae	*Luehea*
Vochysiaceae	*Vochysia*

There is a modified condition of this in *Rhammatophyllum* (Cruciferae), and small, almost rudimentary wings are seen in some Papaveraceae and Fumariaceae and Hypericaceae.

4. *The occurrence of rare characters*

The fourth general example of repetition to be cited is the occurrence among widely differing plants of what can only be called rare conditions. Of this a great many cases might be quoted but the mention of four must suffice to call attention to many of the interesting points.

Sucker tendrils are found in Vitaceae, e.g. *Parthenocissus quinquefolia;* in Bignoniaceae, e.g. *Pithecoctenium phaseoloides;* and in Cucurbitaceae, e.g. *Peponopsis adhaerens.*

A less definitive but no less noteworthy resemblance is seen in those plants, various in detail, which bear their flowers at or even below ground level, among these being *Pitcairnia corallina* (Bromeliaceae); *Paraphyadanthe* (Flacourtiaceae); *Agalmyla staminea* (Gesneriaceae); *Gomphostemma chinense* (Labiatae); *Geanthus* (Liliaceae); *Disteganthus basilateralis* and *Geogenanthus* (Commelinaceae); the strange orchid *Rhizanthella*; *Stylochiton* (Araceae); and a species of *Juelia* (Balanophoraceae).

A third instance is afforded by the occurrence of flowers on either the edge or surface of leaf-like organs. The former is well known in part of the genus *Phyllanthus* (Euphorbiaceae), and is seen also in *Semele* (Ruscaceae) and *Pachynema* (Dilleniaceae), while the condition in *Carmichaelia* and *Bossiaea* (Papilionaceae) and in *Muehlenbeckia platyclados* (Polygonaceae) is not very different. The latter is familiar in *Ruscus* (Ruscaceae) and is to be seen also in *Helwingia* (Cornaceae), *Phyllonoma* (Saxifragaceae), *Mocquerysia, Phyllobotryum* and *Phylloclinium* (Flacourtiaceae), *Erythrochiton* (Rutaceae), and *Cneorum* (Cneoraceae). The condition in *Polycardia* (Celastraceae) is between the former and the latter.

Still another example of repetition is seen in those plants which have prickly leaves, superficially at least of the holly sort, among them being *Osmanthus ilicifolius* (Oleaceae), *Alchornea ilicifolia* (Euphorbiaceae), *Comocladia ilicifolia* (Anacardiaceae), *Itea ilicifolia* (Escalloniaceae), and *Desfontainia spinosa* (Desfontainiaceae).

Superficial Resemblance

Although there is seldom any satisfactory way of discovering how or where particular characters have originated, it may be imagined that repetition as just described is, as often as not, a consequence of differentiation, that is to say, it results from the appearance within a lineage of some condition not hitherto expressed therein. Moreover, judging not only from what has been said, but also from general experience, it is apparent enough that characters thus subsequently acquired are, or become, more or less independent of others, and thus may contribute to a great many combinations. Indeed, it is this more than any other circumstance which accounts for the astonishing multifariousness of the Flowering Plants. But this is only one possible result of repetition. It may equally well contribute to homogeneity rather than heterogeneity, because the appearance of a new factor in a lineage may, though adding something of diversity within that lineage, lessen the visual gulf between that lineage and others. It is when repetition tends to this result that it becomes the basis of what is, more especially, called *superficial resemblance*.

As has been said, superficial resemblances are usually defined as likenesses that are not the immediate consequence of close blood relationship, but rather coincidental, and that is what is meant by the term here. They may also be thought of as analogies rather than homologies. The word superficial applies to them in its proper sense, and it is regrettable that it has also acquired a certain flavour of contempt, because the problems of the superficial resemblances between living things are as urgent as any others.

The close connection between repetition and superficial resemblance cannot be better stressed than by saying that though repetition need not necessarily result in superficial resemblance, superficial resemblance itself cannot occur unless characters are, in some way and to some extent, repeated, and it is therefore not surprising to find that there are many different expressions of such resemblance according to the kinds of characters contributing to the visual effects produced. No doubt an elaborate classification of these could be made but there is followed here the simpler plan of recognizing four broad sorts of superficial resemblance, and of describing them by referring to certain selected examples of each. First there are the cases in which superficial resemblance is a matter of some or all the vegetative parts only and in which it becomes lessened or even destroyed when the plants are in flower and fruit. Second is the sort in which resemblance resides especially in the flowers, which for this reason are usually of similar biological mechanism. In the third sort the resemblance is not only in the vegetative parts but also in the flowers, so that a wider similarity results, and it is to this sort that the generalized conception of superficial resemblance most naturally and typically applies. In the fourth sort, which is really a more particular version of the second and, to a less degree, of the third, one or sometimes both the contributors to the resemblance is a pseudanthial plant, usually a Composite. This fourfold arrangement is useful, but it must not be over-stressed nor allowed to obscure the fact that superficial resemblances are extremely varied in almost all respects. Hardly any two examples can be exactly equated and almost every one has something of peculiar interest about it.

One intriguing aspect of superficial resemblance which merits a moment's thought is its relation to the differences between plants and animals discussed in Chapter 2. In animals many, and certainly some of the most striking, of such similarities are connected with volitional movements of various sorts and more especially with those involved in the pursuit and ingestion of food, and so they particularly reflect the essentially active mode of animal life. In plants, on the other hand, nutritional processes are, comparatively, so passive that they call for less complex apparatus, and the whole paraphernalia of feeding, structure and behaviour in active living animals finds its counterpart among plants in the production of expanded chlorophyll-containing organs. It might even be argued that, since leaf form seems to have little or no bearing on blood relationship, the resemblances between plants in this respect are in total superficial by definition, and that this in fact provides one of the most widespread of such similarities in the Flowering Plants. Again, it is probably true to say that in animals the reproductive phases afford relatively few examples of superficial resemblance. No doubt this is mainly because similarities in reproductive characters are for the most part regarded as indicative of kinship and are therefore not, by definition, superficial, but it would seem also partly due to the fact that, in comparison with plants, animals are in some senses at their least active during reproduction. In plants it may be held that the reverse is true and that their most active life phases, in the sense of the least completely static, are the two most directly associated with reproduction, namely flowering, which after all is largely only the paraphernalia of pollination, and fruiting, which is similarly only the paraphernalia of dissemination. Whether this is a fair comparison or not resemblances involving these phases of plant life are notably frequent.

The great difficulty in writing about superficial resemblance is, of course, how satisfactorily to convey in words to the reader, likenesses which cannot, by their very nature, and almost by their very definition, be susceptible to ordinary methods of description and comparative estimation. Likeness of this kind is essentially a matter of visual perception, and what in the eyes of one beholder may be a striking resemblance, may to another amount to something much less than this, or even it may be, to something so debatable as to be denied any real existence at all. Moreover, it is very difficult to be wholly objective in conveying observed effects of this sort, and especially to avoid using tendencious words which may because of their common meanings imply something that may prejudge the issue. Nor do pictures help in this as much as might at first be expected, for they can, just as easily and unintentionally, be no less tendencious than the written word. So much, too, depends on colour that nothing short of the highest techniques of colour reproduction and draughtsmanship is really satisfactory, and this is equally or even more true of photography. Partly because of these practical difficulties, but partly also to avoid as much as possible the danger of influencing the reader's appreciation of the various cases cited, no attempt has been made, either directly or indirectly, to illustrate this chapter pictorially. There is no substitute for direct observation of the living plants but many of those mentioned are familiar or illustrated in some easily available work such as one or other of those mentioned in the Preface, and for the rest pictorial sources should be sought in the *Iconum Botanicarum Index Londinensis,* London, 1929-31.

Examples of superficial resemblance largely or entirely involving vegetative organs only

1. A good instance from the border line between mere repetition and truer superficial resemblance is seen in the cactoid Flowering Plants, in which the stems present that characteristic form of succulence seen in many columnar Cacti. Such plants constitute three taxa which are very dissimilar in other respects, namely the American family Cactaceae; the Old World and almost exclusively African subtribe of the Asclepiadaceae the Stapelieae (see Chapters 7 and 8); and a predominantly African portion of the great genus *Euphorbia* (Euphorbiaceae). Between these three there is a most interesting combination of general similarity and detailed differences.

2. For many years the Australian genus *Byblis* and the South African genus *Roridula* were included in the widespread family Droseraceae because their vegetative parts were so like those of many of the latter that they were supposed to have a similar insectivorous mode of nutrition. It is now held that their leaves have no digestive powers and they are made into one, or two, separate small families on the more usual basis of certain floral characters. On this interpretation the appearance of the vegetative parts gives them a remarkable superficial resemblance to members of the Droseraceae which is not much lessened when they are in flower.

3. The Eurasian plant hemp (*Cannabis sativa*) belonging to the Cannabinaceae, and the Asiatic plant *Datisca cannabinum* (Datiscaceae) share an almost unique general vegetative design which is strongly emphasized by the simple dioecious flowers. This is however a particularly good instance of superficial resemblance because the leaves of the two plants are in fact palmately and pinnately dissected respectively.

4. Certain species of the series Sibiricae of the genus *Saxifraga* (Saxifragaceae), and especially *S. cernua,* have a characteristic general facies in which length of petiole, shape of leaf blade, presence of bulbils and general inflorescence form are perhaps the chief components, and these are almost exactly repeated in

the Alaskan plant *Romanzoffia sitchensis* (Polemoniaceae), and the effect is enhanced by the flowers, although those of the latter have a corolla tube and several other distinctive features.

Examples of superficial resemblance residing chiefly or entirely in the flowers

5. A striking but rather borderline case since it really involves only a single somewhat unusual character is that of the fringed flowers of the Alpine genus *Soldanella* (Primulaceae) and the Japanese genus *Schizocodon* (Diapensiaceae). The plants have rather similar vegetative features but differ in flower colour.

6. To illustrate how much alike flowers may appear to the casual observer despite great differences in detailed structure there may be cited two plants not uncommonly seen growing near one another in glasshouses, the African *Impatiens sultani* (Balsaminaceae) and the Brazilian *Brunfelsia calycina* (Solanaceae). Both in outline and in hue the flowers, when seen amongst the foliage, are much the same, but while the first has no corolla tube and a long spur, the latter has a long tube and a slightly irregular limb.

7. Certain genera, of which *Ribes* (Grossulariaceae) is a good example, are remarkable for the great variation of flower form among the species. This often results in superficial resemblance, and two instances in particular may be mentioned. The first is the notable general resemblance, chiefly expressed in the flowers, between species of this northern and Andean genus *Ribes* and the genus *Neillia* (Rosaceae) which comes from Asia and Malaysia. The second is that of the North American species *Ribes speciosum*, in which the flowers are more or less solitary, though numerous, fully pendent with exserted stamens and styles, and therefore have much in common with some members of the chiefly American genus *Fuchsia* (Onagraceae). This resemblance is essentially a functional one only and is anything but enhanced by the vegetative characters. It may be added that the flowers of at least one Monocotyledon, *Pitcairnia staminea* (Bromeliaceae) are also similar.

8. This is of special interest because it concerns plants often found growing in proximity in this country under natural conditions, namely *Cardaria* (*Lepidium*) *draba* (Cruciferae) and certain members of the Umbelliferae, especially perhaps *Anthriscus*

(*Chaerophyllum*) *sylvestris*. Its chief significance lies in the fact that the former is atypical of its family in certain respects which cause it easily to be mistaken superficially for a member of the latter, notably in having markedly corymbose racemes in which at early anthesis all the flowers tend to be at the same level and thus to simulate an umbel, and in the detailed designs of the individual flowers. The characteristic "lace" effect produced by so many Umbellifers when their flowers are in mass results not only from their mutual dispositions but also because each flower is, in effect, a shallow cup or basket of somewhat upturned petals which are incompletely contiguous and between which are the spreading stamens. In most Crucifers the flower has, on the contrary, a more or less deeply cylindrical calyx and the stamens are closely grouped, but in *Cardaria draba* not only is the flower shallow and open, but the posture of the petals and stamens is much more like that of Umbellifers. Hence the resemblance is due largely to the fact that the *Cardaria* is, among Crucifers, anomalous in particular ways which help to bridge the normal visual gaps between the two families.

Examples of superficial resemblance to which both vegetative and floral characters contribute

Resemblances *within* either the Monocotyledons or Dicotyledons.

(*a*) Resemblances in which neither contributor is particularly anomalous for its kind.

9. This sort of resemblance is well introduced by one of the most familiar and complete of all examples, namely that between certain species of the very widespread genus *Ranunculus* (Ranunculaceae) and certain species of the widely northern genus *Potentilla* (Rosaceae). It is unnecessary here to detail all the points but they extend to so many aspects of the plants that these are usually held to be separable only by the presence of stipules and of an epicalyx in the latter.

This example also illustrates the problem of consanguinity which has always to be taken into account. The two families here concerned are, certainly, not very far apart genealogically but the superficial resemblance described can hardly be accounted for solely on this basis.

10. There is a remarkable over-all similarity of appearance, involving both vegetative parts and flower colour and arrangement, between the Australian *Brunonia australis* (Brunoniaceae), the European *Globularia nudicaulis* (Globulariaceae) and the European *Jasione perennis* (Campanulaceae), though the floral details are widely different.

11. A still more multiple superficial resemblance is afforded by the various plants which are small herbs with basal rosettes of small entire sessile leaves and single peduncles bearing a more or less bracteate simple umbel or umbellate cluster of small regular flowers at the top. The widely distributed genus *Androsace* (Primulaceae) is perhaps the most typical here, but the North American genus *Abronia* (Nyctaginaceae); the North American genus *Eriogonum* (Polygonaceae); the northern and Andean genus *Saxifraga* (Saxifragaceae); the chiefly American genus *Verbena* (Verbenaceae); and doubtless various others include very similar plants, though the flower colour is rather variable.

12. This is a special case involving the remarkable North temperate parasitic genus *Arceuthobium* (Loranthaceae) on the one hand, and the Old World tropical genus *Apodytes* and the tropical American and African genus *Pilostyles* on the other, both the latter (also parasites) belonging to the Rafflesiaceae. All these live in woody hosts and betray their presence only when their small stalkless flowers emerge from the superficial tissues. The details of these flowers are, however, very different so that the superficial resemblance is one of highly specialized biology rather than anything else. At the same time their form is not the only and inevitable consequence of their habit.

(b) Resemblances in which one contributor is distinctly anomalous for its kind.

13. A particularly remarkable superficial resemblance is that between such members of the genus *Delphinium* (Ranunculaceae) as the Siberian *D. elatum* familiar in herbaceous borders, and the tropical African *Coleus thyrsoideus* (Labiatae) not uncommonly seen in glasshouses, though they flower at different times of the year. Here the resemblance is superficial in the purest, visual, sense of the word, for the two plants, so like one another at first glance are soon seen to be very different. The important point

is that this apparent resemblance is caused by the fact that the *Coleus* is unlike the usual run of Labiates in certain specific ways which make it look more like the *Delphinium*. It has round stems and the typical rather pale green foliage of *Delphinium* and, in addition, its leaves are irregularly toothed in a way distinctly suggestive of the foliage of that plant. The inflorescence, also, is a fairly long raceme of comparatively long-stalked partial cymes, in each of which the flowers are so grouped as to form units very similar in size and colour to *Delphinium* flowers, while even within these cymes the increased shadow towards the centre simulates that of the *Delphinium* flower. In short, at first sight these two plants give almost the same visual impression; it is interesting to speculate whether it appears so also to potential insect visitors.

Resemblances *between* Monocotyledons and Dicotyledons.

14. Another superficial resemblance which can easily be assessed in the field is that between the European *Plantago maritima* (Plantaginaceae) and the widely temperate *Triglochin maritima* (Juncaginaceae), in which there is not only the most complete morphological resemblance but also a close parallel ecologically, both plants often growing together in salt marshes. The structural likeness springs chiefly from the foliage of the former which is remarkably like that, more typical, of the latter, but it extends also to general habit and silhouette, the last being contributed to chiefly, perhaps, by the long plantain-like spikes of the *Triglochin*. Thus, both partners contribute but the *Plantago* may be regarded as the more anomalous since its leaf form is very uncommon in Dicotyledons. Indeed, as other examples will show, superficial resemblance *between* Monocotyledons and Dicotyledons generally rests on anomalous foliage in the latter.

15. Another very striking over-all resemblance is that between two plants of the European mountains, *Gentiana lutea* (Gentianaceae) and *Veratrum album* (Liliaceae). Here neither plant is entirely typical of its kind, the former having, in its smoothness and leaf-venation, a distinctly monocotyledonous facies, and the latter in breadth of leaf and general design of flower much the appearance of a Dicotyledon. Stature and colour are also on the whole the same though the flowers are more conspicuous in the former. The familiarity of these plants adds to their interest but in

fact there is an even more remarkable similar likeness between the Chinese *Veratrilla bailloni* (Gentianaceae) and other species of the genus *Veratrum*.

16. To those familiar only with the foxglove it is a surprise to find that some others of the genus *Digitalis* (Scrophulariaceae) have strong resemblance to certain orchids. This is especially the case with the eastern European species *D. laevigata* and the Levant species *D. orientalis* which have a peculiar likeness to such European orchids as *Epipactis palustris* in particular and certain species of *Cephalanthera*. Not only are their leaves narrow, smooth and almost parallel-veined but the upper have the curved profile so well seen in many Monocotyledons and are set widely apart on smooth stems. The flowers also have much the spacing and posture of the orchids, are strongly bilabiate, and, most impressive, have a mixture of colours reminiscent of some of those plants. In smaller, fewer-flowered specimens of the *Digitalis* species the general similarity of appearance is most remarkable.

17. Rather similar, in general terms, to the last is the strong likeness between some members of the section Orchidioides of the world-wide genus *Utricularia* (Lentibulariaceae) and certain orchids, more especially the two American species *Ionopsis utricularioides* and *Comparretia falcata*. Resemblance here is over-all but resides chiefly in the strongly zygomorphic, spurred, bilabiate floral design (enhanced by the fact that *Utricularia* has an andrecium of only two unilocular anthers) and in the shape and texture of the basal rosette of leaves. With regard to this last point, it is interesting to recall that the photosynthetic blades of *Utricularia* may not be altogether homologous with ordinary dicotyledonous leaves.

18. Reference was made in Chapter 3 to the fact that some of the most real distinctions between Monocotyledons and Dicotyledons are both subtle and hard to define, and this is well illustrated here by certain species of the tropical American genus *Aphelandra* (Acanthaceae) which, because they possess some of these features in remarkable degree, have a strong likeness, particularly in flowers and in foliage respectively to the American tropical monocotyledonous genus *Canna* (Cannaceae) and the almost entirely tropical American monocotyledonous genus *Calathea* (Marantaceae).

19. It will be apparent that one of the most potent factors in resemblance between Monocotyledons and Dicotyledons is the possession, by the latter, of Monocotyledon-like leaves, and there are many other instances of this, among which there may particularly be mentioned the likeness between species of the Australian genus *Dracophyllum* (Epacridaceae) and the austral genus *Astelia* (Liliaceae) which extends in some species to the inflorescences also, and the case of the anomalous Brazilian Composite, *Schlechtendalia luzulaefolius* which, vegetatively at least, is reminiscent of many more generalized Monocotyledons. It is not irrelevant to note that what appear to be "phyllodinous" leaves are not infrequent in the Epacridaceae, and occur notably also in the tribe Thibaudieae of the Vacciniaceae.

Examples of superficial resemblance in which at least one contributor is a pseudanthial plant

Resemblances in which members of the Compositae simulate euanthial plants.

20. One of the most generally familiar instances of superficial resemblances of this sort is that between *Achillea millefolium* (Compositae) and various members of the Umbelliferae, but this is in fact a somewhat crude example and certain deeper pink forms of *A. millefolium* are much more like similarly coloured forms of *Spiraea japonica* (Rosaceae). In both cases, however, the likeness is confined to the inflorescence.

21. As explained in Chapter 10 various Composites with five-rayed capitula more or less closely resemble euanthial plants and three may, in particular, serve as instances here. In the North American genus *Chrysogonum* the detailed form, colour and distribution of the capitula (and notably the shape of the rays) make them very like ordinary yellow pentamerous flowers of medium size. The small Andean white-rayed *Nassauvia bryoides* is very like a small mossy saxifrage: and the Chilian blue-flowered *Triptilion spinosum* is very reminiscent of the North American *Gilia pungens* (Polemoniaceae).

22. Of a rather different kind is the curious resemblance not infrequently to be observed in gardens between some cultivars of *Tagetes patula* and some colour conditions of *Tropaeolum majus*

(Tropaeolaceae), both American plants. The number of rays in the Composite is five and they have the same shape, colour and size as the petals of the *Tropaeolum*, while in both there are large basal patches of reddish brown.

23. A much more specialized resemblance is that between the Australian Composite genus *Humea*, often grown as the "incense plant", and certain members of the Amaranthaceae. The latter are characterized by highly multiple inflorescences, which may be more or less pendent, of innumerable tiny scarious flowers, and the general overall appearance of such inflorescences is closely reproduced by the semi-pendent compound inflorescences of minute, and actually one-flowered, capitula of *Humea*, which in consequence is very unlike others of its own family.

24. The most aberrant Composites are among those in which there are compound capitula and this is strikingly confirmed by the odd superficial resemblances which sometimes result. A remarkable instance is that between certain species of the Australian Composite genus *Angianthus*, notably *A. myosuroides*, and the temperate genus *Myosurus* in Ranunculaceae.

Resemblances in which euanthial plants simulate Composites.

25. The most general instance of this, the similarity between Composites and some of the Ranunculaceae in which there are more tepals than usual has been touched upon at some length in Chapter 10, and it will suffice here to mention another good example of it namely that between species of *Adonis* and *Atragene*, such as the South African *Atragene capensis* (Ranunculaceae) and such Composites as the S. African and Australian genus *Arctotis*.

26. A very unusual example of superficial resemblance is that involving the Australian genus *Dipteranthemum* (Amaranthaceae). Here the flowers are in heads subtended by a small series of overlapping bracts but the main interest is in the calyces, in each of which two of the more or less erect bluntly linear lobes are much larger and longer than the rest. Moreover, these are described as orange or apricot on the exposed (under) surface, and mother-of-pearl on the less obvious upper surface. Because of these calyces there is a strong general resemblance to various Composites while the colour repeats the scheme of at least one species of the composite genus *Arctotis*. It is interesting that *Dipteranthemum* is

closely related to the genus *Triclinium* in which however the flowers are in short thick spikes so that their enlarged calyx lobes are not in a single series and therefore give no illusion of Composite rays.

27. In the Philippine Orchid *Cirrhopetalum nutans* the lips of the flowers are of such a shape and posture that the whole inflorescence looks, at first sight, very like a large, nodding pale-yellow Composite capitulum, though the absence of anything corresponding to the involucre is soon perceived.

Other resemblances involving pseudanthial plants.

28. It is not, perhaps, surprising that there is close resemblance between some of the least particularized Composites and some of the more highly subpseudanthial or even quite pseudanthial plants in other families. A good instance of this is the Australian *Ptilotus helichrysoides* (Amaranthaceae) which, as its name indicates is like one of the helichrysoid Composites. Much the same is true also of the S. African genus *Staavia* (Bruniaceae). A slightly different condition is that of some species of the mainly South African genus *Phylica* (Rhamnaceae) in which there is a slight parallel with such Compositae as *Leontopodium*.

29. A remarkable pseudanthial resemblance is that between certain orchids of the tropical American genus *Masdevallia*, notably *M. ignea*, and such aroids as the tropical American *Anthurium scherzerianum*. In the former, which is the anomalous contributor, the lower lip, though somewhat divided, has the colour and much of the shape of the flat spathe of the latter, while the upper part of the orchid flower has the form of a yellow finger reminiscent of the spadix of the aroid. The foliage and general stature of the orchid enhances the likeness to an Aroid.

30. Strict superficial resemblances between two strongly pseudanthial plants appear to be very unusual and the following four cases may suffice to illustrate them. The first is that of such species of *Cornus* as the North American *C. florida* (Cornaceae) (fig. 138) and the Himalayan *Parrotiopsis jacquemontiana* (Hamamelidaceae); the second that of such Proteaceous plants as *Dryandra* and some Compositae like *Carlina* and *Stifftia* respectively; and the third is that of *Pimelea physodes* and the kind of Protead exemplified by *Mimetes zeyheri*, but none of these is very arresting. In the first some consanguinity may be involved,

in the second and third the resemblance is rather the inevitable result of pseudanthial specialization, though of course these are of no less interest on that account.

The fourth instance is rather different since it involves two members of the same family (Compositae) and may therefore be associated in some measure with true relationship, but it is so interesting that it certainly merits mention here. Those with a general knowledge of the Composites are likely to agree that one of the most characteristic vegetative designs among them is that shown by such genera of the Liguliflorae as *Sonchus*. It is remarkable that this is almost exactly reproduced in certain species of the genus *Emilia* (e.g. *E. sonchifolia*), which is often regarded as a tropical subgroup of the great genus *Senecio*, which belongs to the Tubuliflorae.

Further examples of superficial resemblance

The selected examples just cited make it difficult to resist the conclusion that some families and genera of flowering plants are more often contributors to superficial resemblance than others and there is much else to confirm this impression.

The family Gentianaceae is notably prominent. One example in it has already been mentioned on p. 344 and there are others very similar, among them *Swertia kingii*, which though much larger-flowered has an equally monocotyledonous appearance. Species of *Canscora* and *Curtia*, despite their floral irregularity, strongly recall some of the gypsophiloid Caryophyllaceae in general appearance. Species of *Centaurium* have been compared with *Viviania* in Geraniaceae, and species of *Pleurogyne* with species of *Parnassia* in Saxifragaceae. Rather more particular parallels are those between some purple-flowered species of *Exacum* and *Chironia* which show andrecial irregularity and characteristic venation, and various members of the Melastomataceae, and between the five-spurred flowers of *Halenia* and those of *Epimedium* or of some kinds of *Aquilegia*. The small genus *Prepusa* shows superficial resemblances in two directions, between *Prepusa connata* and such species of *Primula* as *Primula agleniana* and between *Prepusa hookeriana* and some silenoid Caryophyllaceae with reddish inflated calyces and white petals. *Sabbatia chloroides*, and perhaps others with more than the usual number of corolla lobes have much the appearance of some Composites of the smaller *Cosmos* sort. The family is also concerned in an interesting reverse or complementary pair of resemblances. In one, *Grammanthes gentianoides*, which belongs to the Crassulaceae, has closely the general appearance, though less the flower colour, of a small species of gentian: in the other the gentianaceous genus *Anagallidium* (founded on *Swertia dichotoma*) has a remarkable likeness to plants of the *Anagallis arvensis* sort in Primulaceae.

Noteworthy here also is the family Orchidaceae, already mentioned more than once. Not only is the general parallel between the flowers of Orchids and those of Balsams familiar, but there are other more particular examples too, as in *Curculigo orchioides* and *Roscoea cautleoides*, both of the Zingiberaceae, and the section *gandavensis* of *Gladiolus* (Iridaceae), all of which recall orchids of some kind rather than others of their own congeners. On the other hand there are species of *Oberonia* (Orchidaceae) which are much like some kinds of *Plantago* (Plantaginaceae) or even of species of the Piperaceae; and species of *Arpophyllum* and *Habenaria* (Orchidaceae) which are very similar to some species of *Polygonum* (Polygonaceae).

Several members of the Cruciferae, among them *Pegaeophyton sinense*, *Ionopsidium acaule* and *Leavenworthia stylosa*, have solitary or apparently solitary rather short-stalked flowers amid a basal rosette of leaves and thus at first sight look much more like various other dicotyledons rather than the rest of their own family. The section Streptantheae of the family has even more definite resemblances, as in *Icianthus hyacinthoides*, which is very like a hyacinth or *Dipcadi*, and in *Microsemia polygaloides*, which has much the appearance of a *Polygala*.

Caryophyllaceae, too, bulk large in superficial resemblances. Besides those mentioned already some members of the genera *Saponaria*, *Silene* and *Stellaria* have a good deal in common with the genus *Phlox* (Polemoniaceae); *Zaluzianskia lychnoidea* (Scrophulariaceae) resembles species of *Lychnis*; and the genus *Buffonia* is often compared with *Juncus bufonius* (Juncaceae). Certain of the Composite genus *Prenanthes* with greenish white rays have much resemblance to such species of *Silene* as *S. saxatilis*.

Nor is it only the more generalized families which are involved in this fashion. Even so well-defined a group as the Labiatae reveals a number of instances. Between Labiatae and members of the Acanthaceae there are various parallels, as in *Acanthomirta ilicifolia* (Labiatae) with its acanthoid-like spiny bracts; in the general appearance of *Salvia involucrata*; and, in the reverse direction, *Brillantaisia lamium* (Acanthaceae) which is one of several with Labiate-like flowers. Other Labiates, because they lack the usual characters of that family, look more like plants of the family Scrophulariaceae, as is well illustrated by *Chelonopsis moschata*, which resembles *Chelone*; species of *Hemiandra*, which are like some of the Gerardieae; and *Prostanthera lasianthus* which in inflorescence at least is very like *Catalpa*.

Though not quite parallel with the groups just set out the genus *Euphorbia* is of considerable interest in respect of superficial resemblance. In most species the greenish yellow colour of the cyathia tends to betray their peculiar pseudanthial organization but in some there are associated features which much enhance a likeness to euanthial plants. The form and brilliant colour of the cyathial lobes or glands in *E. jacquiniflora* (fig. 140) recalls some of the Rubiaceae; while the presence of only three glands in *E. hystrix* suggests a monocotyledonous corolla. Most notable perhaps is *E. petiolaris* for here the whole plant is a small semi-procumbent, or at least weak, herb with widely spaced leaves in the axils of which arise the solitary, long stalked, yellow cyathia, and an over-all appearance reminiscent of that of some Aizoaceae, Molluginaceae or Portulacaceae.

Finally, we may revert for a moment to a matter mentioned much earlier in the chapter, namely that of the repetition of rare characters, for it is easier in the light of what has since been said to see the relation between this and superficial resemblance. Often if the character is a very peculiar one giving a decided impress to its possessor the distinction between repetition and superficial resemblance wears very thin, as is illustrated for example in the two cases of *Cyclanthera* (Cucurbitaceae) and *Phyllanthus cyclanthera* in Euphorbiaceae, in each of which there are annular anthers, and of *Caltha dionaeifolia* (Ranunculaceae) and *Dionaea muscipula* (Droseraceae) with their strange leaves. Especially is this so where the rare character contributes to a comparatively common condition of a different kind and here the genus *Antonia* of Loganiaceae is interesting. *Antonia* exhibits the rare character of a multi-seriate or involucrate calyx, but an analogous structure is characteristic of Composites, with the result that, assisted by some rather general vegetative similarities, *Antonia* is superficially, and especially at certain states of the inflorescence, very much like some of the plucheoid or eupatorioid Composites.

Such are some of the more interesting or, from the present point of view, important, instances, of superficial resemblance. They do not, however, by any means exhaust the subject and since some readers may wish to know of still other cases this chapter may well conclude with a short further list couched in brief form. In some, resemblance is wide, and in others it is restricted to flowers or fruit, but all have something to contribute to the general problem:

Anabaenella (Euphorbiaceae): *Dioscorea* spp. (Dioscoreaceae).

Thomasia solanacea (Sterculiaceae): *Solanum nelsoni* (Solanaceae).

Symplocos crataegoides (Symplocaceae): *Crataegus* spp. (Rosaceae).

Dasypogon bromeliaefolius (Liliaceae): *Thurnia sphaerocephala* (Thurniaceae).

Microlicia spp. (Melastomataceae): *Roella* spp. (Campanulaceae).

Pfaffia spp. (Amaranthaceae): *Trifolium* spp. aff. *arvense* (Papilionaceae).

Rhipsalis cassytha (Cactaceae): *Cassytha* spp. (Lauraceae).

Leucogenes leontopodium (Compositae): *Leontopodium* spp. (Compositae).

Pterisanthes spp. (Vitaceae): *Dorstenia* spp. (Urticaceae).

Lactoris (Lactoridaceae): *Buxus* spp. (Buxaceae).

Billbergia spp. (Bromeliaceae): *Riedelia* spp. (Zingiberaceae).

Haworthia spp. (Liliaceae): *Cotyledon* and *Sempervivum* spp. (Crassulaceae).

Phyllacantha grisebachiana (Rubiaceae): *Colletia cruciata* (Rhamnaceae).

Peperomia spp. (Piperaceae): *Pothos* spp. (Araceae).

Mesembryanthemum brownii etc. (Aizoaceae): *Aster* spp. (Compositae).

Grayia polygaloides (Chenopodiaceae): *Polygala* spp. (Polygalaceae).

Steriphoma paradoxum (Capparidaceae): *Caesalpinia gilliesii* (Caesalpiniaceae).

Actinidia kolomikta (Actinidiaceae): *Prunus cerasus* (Rosaceae).

Oxypetalum strictum (Asclepiadaceae): *Solanum* spp. (Solanaceae).

Antholyza spp. (Liliaceae): *Leonotis leonurus* (Labiatae).

Clematis hieracifolia (Ranunculaceae): *Hyacinthus* spp. (Liliaceae).

Cleome spp. (Capparidaceae): *Bauhinia* spp. (Papilionaceae).
Amaranthus chlorostachys (Amaranthaceae): *Setaria italica* (Gramineae).
Musschia aurea (Campanulaceae): spp. aff. *Lilium* (Liliaceae).
Siphocampylus manettiaeflorus (Lobeliaceae): *Manettia ignita* (Rubiaceae).
Isotoma longiflora (Lobeliaceae): *Clerodendron* spp. (Verbenaceae).
Isocarpha achyranthus (Compositae): *Achyranthes* spp. (Amaranthaceae).
Hymenoclea salsola (Compositae): *Salsola* spp. (Chenopodiaceae)

NOTE ON LITERATURE

References relating directly to the subject matter of this chapter appear to be few, but the following may enable the reader to pursue some aspects of it a little further.

Arber, A.: *Monocotyledons*. Cambridge, 1925. Especially the final chapter.

Bennett, A.: "Mimicry in Plants." *Popular Science Review*, **11**, 1872.

Cooke, M. C.: *Freaks and Marvels of Plant Life*. London, 1882.

Darwin, F., ed. : *Life and Letters of Charles Darwin*, vol. III, p. 70. London, 1887.

Hildebrand, F.: *Uber ahnlickeiten im Pflanzenreich*. Leipzig, 1902.

Marloth, R.: "Mimicry among Plants." *Trans. S. Afr. Phil. Soc.*, **15**, 1904.

Marloth, R.: "Further observations of mimicry among plants." *Trans. S. Afr. Phil. Soc.*, **16**, 1905.

Robson, G. C. and Richards, O. W.: *The Variation of Animals in Nature* (first chapter). London, 1936.

Saunders, W. (reports of an exhibit by): *Nature*, May 26, 1870 and May 4, 1871.

Wallace, A. R.: *Tropical Nature and Other Essays*. London, 1878. Especially chapter six.

CHAPTER TWELVE

SUMMARY AND CONCLUSIONS

THE INTRODUCTORY CHAPTER presented an argument which, in the briefest terms, may be reiterated as follows:

1. The present state of the scientific study of evolution is unsatisfactory because many of the ideas and hypotheses used in it are derived too exclusively from the study of certain kinds of organisms only.

2. There are reasons of several sorts for this, but the practical result is that there has been a lack of balance in evolutionary investigations, some aspects of the matter having been unduly stressed and others unduly ignored. Owing to this generalizations about evolution are often based on an insufficient appreciation of all the relevant facts.

3. The remedy for this is to test the validity of these generalizations and of other concepts against additional and less familiar facts by broadening the scope of evolutionary studies to include those fields of enquiry which have so far received less than enough attention.

4. Among these relatively neglected fields two, especially, stand out, the Flowering Plants and the Insects. Both have features of great and particular potential interest in relation to evolution, and the former have the added significance that they occupy a unique and fundamental place in the general scheme of nature.

5. It is particularly desirable, therefore, and likely to be particularly profitable, to recall some of the vast number of more relevant facts about these plants, and to consider what applications these may have with regard to various evolutionary hypotheses and concepts.

This argument rests, of course, on the premiss that plants, and especially Flowering Plants, are of no less interest to the student of evolution than are animals, and this opinion involves not only the question of the similarities and differences between the two, but that of their biological relationship. It seems likely that the comparative neglect of plants in connection with evolution has come about through an imperfect understanding of the circumstances in these matters, which were therefore treated at some length in a second chapter.

CHAPTER 2 LED TO THE CONCLUSION that the resemblances between plants and animals amount to a community of organizational design which must be regarded as expressing a fundamental unity of origin, and this being so that the more obvious differences of function and form are, by comparison, superficial only, illustrating no more than that there are two distinct ways of making one and the same kind of machine work, or to change the metaphor slightly, illustrating two bodies politic, one with a closed and one with an open system of economy. On this interpretation it seems highly probable that evolution in the two cases will be a reflection of this situation, being in both the same basic process, but expressing itself with a different distribution of emphasis in each.

This alone is good reason for believing that plants are of no less intrinsic interest in any broad approach to evolution than are animals, but there must further be taken into account the peculiar characteristics of plants, and these make it almost inevitable that plants will exhibit many aspects of evolution in relatively simple fashion, and in a way, therefore, which may be highly advantageous to those who try to understand them. Furthermore, the overall relationship between plants and animals in a state of nature is such that the evolution of animals must be regarded as having been, and as continuing to be, conditioned by the evolution of plants, and thus as in some degree consequent upon this.

No matter what the story of its origin may be organic nature has, for what is presumed to be the greater part of biological time, been expressed in two forms, plant and animals, and the relevant facts about these seem clear enough. Higher plants are not only autotrophic but provide in total all but a small fraction of the vegetable resources of the land world. Higher animals are heterotrophic and hence depend, directly or indirectly, on those vegetable resources. In other words, the higher animals are dependent on the higher plants. Moreover, a land stripped of its higher plants would, no matter how varied it might otherwise be, afford neither habitat nor foothold for the higher animals, and thus the presence of the former is not only a necessary precursor of the presence of the latter, but a necessary provision for the latter's support. A relation of this sort has much in common with that between host and parasite, and scientific propriety will not be greatly outraged if the relationship now observable between plants and animals is thought of in this way.

All this being so it seems an inescapable conclusion that evolution in plants has set both the pace and the course for much evolution in animals, though care must be taken to understand exactly what this conclusion means. It does not mean necessarily that every aspect of animal evolution is directly related to the presence of plants. For instance, there has been a similar developmental tendency in both kingdoms towards the increasing equipment and protection of the unborn young, and this would seem to be something over-riding any purely ecological relation. Similarly there are many structural elaborations in animals which cannot reasonably be referred primarily to ecology. On the other hand it is very noteworthy indeed that animal evolution has never, so far as is known, produced an autotrophic animal, and so has always remained in phase, as it were, with plant evolution, never, in its amplitude, seriously distorting the metabolic relation between the two. The available evidence indicates that there has been a succession of different plant types in time, and the evidence indicates that this is true also of animals, but the all-important point is that the general relation between the two has continued to be the same.

That the maintenance of this relation has imposed conditions on animal evolution can scarcely be denied. There must always have been an appropriate ratio between herbivorous and carnivorous types; not always necessarily the same ratio, since this may have varied from time to time, but a just ratio according to the circumstances. Furthermore, as one dominant plant type succeeded another, the nature of the food material, as well as the circumstances of its presentation, must have altered, and there must have been corresponding changes in the herbivorous fauna. To take one of the most prominent instances it is hard to believe that the evolution of the grasses and the evolution of many of the hypsodont or high-crowned-toothed grazing animals did not bear the relation of cause and effect.

On the other hand there cannot easily be detected, in the general evolutionary story, any compensatory conditioning of the evolution of plants by that of animals. There are, it is true, certain aspects of evolution in the Flowering Plants which it is possible to regard as examples of this, notably the insect-flower relation and, less conspicuously, the development of thorns and other so-called protective devices, but even here the matter is by no means self-evident. Moreover, we are speaking here of the grand course of evolution, against which even these examples fade into relative insignificance. The whole picture of evolution as seen in the light of present-day knowledge, is basically one of the gradual and majestic unfolding of plant form, offering in its immediate wake an ever slowly changing potential environment, the exploitation and population of which has been the main theme of animal evolution. *Spartina townsendii* affords a nice example of this. This grass, since its appearance less than a century ago, has here and there, in suitable places hitherto uncolonized by flowering plants, produced a new and distinctive sort of habitat for other living things, and it may be confidently expected, from general experience, that this will in the course of time become inhabited by species of animals peculiar to it.

On this view it becomes almost axiomatic that evolution in plants is a simpler and more direct expression of that great biological phenomenon than is evolution in animals. There is seen in plant evolution on this grand scale something which seems to have unrolled irrespective of, and as it were, unconscious of the animal kingdom. If this is indeed so then there are two corollaries, first that evolution in plants cannot, on any score, be treated as of less importance and interest than animal evolution, and second, that there may well be learnt from it much that cannot so easily be learned from the more complicated evolution of animals.

CHAPTER 3 DEALS WITH WHAT is at once the main expression of evolution in the Flowering Plants and one of its most mysterious features, the development of two huge and similar, though taxonomically unequal, series of plants, the Monocotyledons and the Dicotyledons, and in particular with the nature of the differences between them and between their respective evolutionary potentials. Because the average Monocotyledon is so readily distinguishable from the average Dicotyledon it may have surprised some readers to find how narrow is the gap between the two series. Certain points of difference are generally to be observed but it is a matter of debate as to whether there is any absolute distinction between them, unless of course it is deliberately decided that every plant with some given character shall *ipso facto* belong to one and not to the other. The nearest approach to an absolute distinctive character appears to be the absence, from monocotyledon bundles (and even more especially from *between* the bundles) of any really functional cambial tissue, so that they are "closed" as opposed to "open", but it would be hard to maintain that this is completely and invariably true. It seems, therefore, that evolution can result in two series of organisms quite distinct from one another in general terms but without any hard and fast diagnostic separation, and one important aspect of a comparison of Monocotyledons and Dicotyledons is the light that it throws on the connection, or lack of it, between relationship and characterization, and how necessary it is to avoid becoming ensnared in the opinion that either of these necessarily represents the other.

The nature of the differences between Monocotyledons and Dicotyledons is also of great interest, for they are, to our human estimation, singularly inconsequent. The difference between one or two cotyledons, for instance, is not only the sort of difference that might arise in a variety of circumstances but one which does not appear to have any great biological importance, either functionally or structurally, for, as was seen, "dicotyledons" with only one cotyledon retain their otherwise distinctive features. Again, if the opinion that the commonest kind of monocotyledon leaf is an expanded bladeless petiole is correct, it is difficult to see the reason for its development. Yet this character is the most generally and obviously distinctive of all, although a few Dicotyledons, notably the Plantagos, appear to be similar.

Probably the difference which, in its ultimate effects, is the most profound, is the distribution of the vascular tissue of Monocotyledons in numerous small and scattered bundles, because such an arrangement inevitably precludes the most usual form of secondary thickening. With regard to this there seem to be three possibilities. The Monocotyledons and Dicotyledons may, respectively, have inherited their conditions from their forebears, which means that they are to be regarded as phylogenetically independent lines. The dicotyledon condition may have been derived, once or more often, from that of the Monocotyledons by the *acquisition* of cambial and other features. The monocotyledon condition may have been derived, once or more often, from that of the Dicotyledons by the *loss* of cambial characters. If the first is true then there must be visualized the independent origin, from different ancestral sources, of two groups with the degree of parallelism that has been described. If the second is true then there must be explained, not only how such acquisition may have come about, but the actual possibility of such a thing, with its obviously enormous expansive effects, happening. If the third is true, there must among other things be visualized the loss from a stock of characters the absence of which must inevitably restrict or at least divert the evolutionary possibilities. Each of these deserves careful and serious thought.

All these three considerations bear upon the more ultimate problem of evolutionary potential. However much caution is used in assigning values to characters, it is surely impossible to resist the conclusion that, in modern colloquial phrase, the Dicotyledons have got something that the Monocotyledons have not got, and that, as a result, their evolution has been much more lavish. The Monocotyledons must, it would seem, be described in general terms as *lacking* cambial activity in the ordinary sense; as *lacking*, in large measure, the more complicated types of leaf form; and as *lacking*, as often as not, heterochlamydy of the perianth, and it is not apparent what, in addition to any features shown by the Dicotyledons, this group of the Monocotyledons has, except perhaps such a difference of prevalence as that in geophily. In so far as this is true, the Monocotyledons are a simplification of the Dicotyledons. Whether this simplification represents a higher evolutionary level is one of those problems which, in the light of present knowledge, it is unprofitable to discuss, but there must be allowance for the possibility that this is so. This would suggest that evolution may result in a loss of characters. At all events it seems that in the Monocotyledons and Dicotyledons there are two extraordinarily parallel series of two quite different evolutionary potentials in certain respects. In certain respects only however, for in others, especially in floristic as opposed to vegetative features, their potentials seem much more similar.

This third chapter ends with a consideration of two related facts which, though familiar enough, are not always fully realized, namely that while the Monocotyledons are, in terms of taxonomy, a much smaller series than the Dicotyledons, the geographical distributions of the two are in total practically the same. It is suggested that difficulties arise here only because the word "smaller" is applied to the Monocotyledons and the chapter goes on to discuss how justifiable the use of this word is. In terms of individual plants (and this after all is the only real measure of the size of groups) it may well be doubted whether the Monocotyledons are, in fact, in a minority. Even in the range of structural variability the disparity between the two is scarcely comparable with the taxonomic disparity, and this sort of difference has little connection with geographical distribution. In short the use of the word "smaller" strictly means only that the Monocotyledons cannot be broken down by the ordinary methods of taxonomic analysis as far as can the Dicotyledons, which is merely to say that the Monocotyledons contain fewer species than the Dicotyledons. This in its turn is an expression of the fact that the discontinuities of form on which the operations of taxonomy are based, are, in the Monocotyledons, of a different average dimension. They are both greater and fewer and thus classify into smaller numbers of units. It thus appears that the taxonomic convention which gives to the Monocotyledons only about a quarter of the number of units it gives to the Dicotyledons is little real measure of relative size, and that there is nothing in it to suggest that, biologically and ecologically, the Monocotyledons are of lesser measure than the Dicotyledons. This distorting idea of size being removed it is no source of great surprise that the respective total geographical distributions of the two series are alike. Incidentally, it may be hazarded that this false conception of size has had a far-reaching effect in appearing to favour the view that the Monocotyledons are in some way or other derivatives of the Dicotyledons, and it may be doubted whether so much discussion would ever have centred round the problem of the phylogenetic relation between the two had they been taxonomically more the same size. Indeed, this is a striking example of the diversionary effects of translating facts too closely in the language of taxonomy which is one of the important themes of the next chapter.

CHAPTER 4 IS THE FIRST OF A TRIAD in which the chief purpose is to present the Flowering Plants primarily *as an achievement of those processes which are called evolution*, on the argument that one of the most promising ways of discovering *means* is to gain a full and proper understanding of *ends*. It is in many respects preliminary to the other two and tries, especially, to introduce the Flowering Plants in somewhat novel terms, with the hope that this may help to disperse the mist of taxonomic formality which obscures so much of our view of the organic world. In this there is intended no denigration either of taxonomy itself or of any of its practitioners. Taxonomy is as essential as any other aspect of biological study, but it is well to remember that it is also no more free from limitations, and that if these are ignored there may be serious interference with a real understanding of the facts. This danger is of course always present, but nowhere, perhaps, is it so live as in the study of evolution.

The method of approach to this problem is to develop the proposition that the only real units in the Flowering Plant world are the individual plants, and that the entities which are called species, genera, families and so on are no more than useful but abstract conceptions of the human mind. Naturally, since otherwise they would be meaningless, these terms express the facts up to a point, but this point is far more distant from reality than is usually admitted. But why, it may be asked, if these classificatory methods are so imperfect have they become so integral a part of biology? This is a very important and proper question and the answer to it is that the facts about living organisms are such, both in their kind and in their mutual relationships, as to appear extremely complicated to the human observer, and even, perhaps, in certain respects, to be beyond the communicative power of the human mind. Some simplification is therefore unavoidable if they are to be represented in speech and print, and the few-category type of classification which is the basis of taxonomic practice is only a special kind of simplification which seems particularly to fit this case. Probably all biologists are agreed that of the innumerable collections of individuals to which the names of species and the rest are given, no two are ever precisely alike or exactly equivalent. To express all such differences would not only be an extremely involved undertaking but, more important, almost certainly one which would need powers of expression and of comparative estimation which are not available. In brief, the truth can be expressed and communicated only in simplified form and the foundation of the way in which this is done is the adoption of the fiction that all natural congeries of individuals belong to one or other of a limited number of categories, the members of each of which are to be regarded as theoretically of equal "value". A formula is, as it were, substituted for reality, and from this there constantly arises a risk that the shadow will be confused with the substance.

In these circumstances the question naturally arises whether it is possible to find any method of depiction which will be more satisfactory than that of formal taxonomy, and it is suggested, at any rate for immediate purposes, that there is at hand an analogy which can be of great service. This is the *stellar analogy*. It is based on the thesis that each individual flowering plant may be likened, especially in respect of its physical and genealogical relationships, to a single member of the celestial system, and more particularly of the galactic system, and that the resemblances and mutual kinships of all these individuals can be *represented* and *pictured* with a high degree of reality by the kind of spatial distribution shown by the stars. The importance of this analogy, and it cannot be too clearly stated that it is no more than an analogy, is partly that it brings our eyes to focus on the true ingredients of the biological world, namely the individuals, and second that it shows the nature of what are nowadays called *taxa*, namely that they are the contents within imaginary boundaries surrounding various totalities of individuals, what those contents are depending on personal taxonomic estimation. The all-important elements in the stellar analogy are two, first that the distribution of the real totalities is variously discontinuous, so that all sorts and sizes of more or less coherent groupings or "clusters of stars" are to be observed, and second that the more any totality is resolved the more discrete it tends to appear. It is the reflection of this kind of distribution as one of the fundamental facts of mundane organic matter which, on the one hand, invites and makes possible the kind of classification which is familiar, and on the other makes the stellar analogy so valuable.

Despite these comments on taxonomic practice and the desirability of restricting, as much as may be possible, the use of such conceptions as species, genera and families, it is obvious that these provide a verbal coinage which cannot possibly be dispensed with altogether, and this is demonstrated in the concluding pages of the chapter, which are devoted to a comparative survey of the larger taxa of flowering plants, namely the families. Its main purpose is to illustrate how different the members of a single taxonomic category are seen to be when they are analysed; how, among them, certain sorts tend to prevail; and, in particular, how the stellar analogy can be usefully pursued further in many directions. It also shows once again the all-important basic variability of biology.

CHAPTER 5 CONTINUES THE TRIAD by presenting the Monocotyledons, and reaches the conclusion that in broad terms, and with very few exceptions, all these plants may be regarded as belonging to one or other of about thirty "sorts", these differing among themselves in content, as well as in their degrees of separation and structural isolation from one another. In respect of the latter some of these thirty or so groups are more isolated and some are less so, and in diagrammatic form the whole series may be pictured as a central core or mass of relatively confluent groups and a peripheral set of more distinct groups. Among the latter it is not difficult to detect four main themes of differentiation. These are the "size" theme, where the outstanding feature is unusual stature, as in the Palms and Screw-pines; the "glumaceous" theme, as it may be called, in which the flowers are small and aggregated into glumiferous inflorescences such as are so familiar in the Grasses and Sedges; the theme of "floral specialization", as is so notable in the Orchids, Gingers and Aroids; and the "aquatic" theme of ecological specialization which finds its extreme expression in the Duckweeds and marine angiosperms. Among the groups of the central core it is similarly possible to see that some, for instance the Spiderworts and the Pipeworts, are more peripheral than others, though on a comparatively small scale of isolation. The very centre consists of a number of almost or completely confluent groups, largely representing the older taxonomic family triad—Liliaceae, Amaryllidaceae and Iridaceae.

This relatively simple arrangement is susceptible without undue difficulty not only to diagrammatic but also to a simple three-dimensional representation in the form of a model and both these are illustrated (figs. 30, 31).

CHAPTER 6 CONVEYS ITS MAIN POINT by its very length, which shows that the Dicotyledons are a much more complex series of plants than are the Monocotyledons. The recognition of the constituent sorts of Monocotyledons was a fairly simple matter but among the Dicotyledons it is far otherwise, and indeed only by modifying our analysis to the extent of speaking of generalized representative types rather than of well-rounded groups can anything useful be done within a limited space. It would seem that the explanation of this greater difficulty is not so much that there are more distinct sorts of Dicotyledons, although in some measure this is certainly true, but that the sorts flow towards and into one another much more. In short the discontinuities within the Dicotyledons are both smaller and more numerous than they are in the Monocotyledons.

One consequence of this is that it is possible to point, in a way which was impossible with the Monocotyledons, to what may be called the generalized type of Dicotyledon, that is to say one from which notable specialization is absent and in which there are medium values in many or all of the main variation gamuts in the series. Not only can this average Dicotyledon be arrived at as an abstraction, by a process akin to that in which a composite portrait may be obtained by superposing many photographs, but it is apparent that the types so pictured are in fact strongly represented among the Dicotyledons. Indeed it is perhaps the most important feature in the constitution of this great assemblage that, to a very notable degree, all the more distinctive and peripheral sorts are connected by others more intermediate in character to a great and close-knit central mass.

But although this sounds quite simple it must be remembered that the differences between the various types are almost infinitely diverse, both in quality and quantity, so that there is again seen here, in rather more particular guise, just that necessity for further simplification and categorization which was discussed earlier and which is the basis of taxonomy in general. In short if the subject is to be both briefly and intelligibly stated even more severe categorization is required.

First of all the number of possibilities must be drastically reduced by combining together various detailed characters into a small number of overall features such as are, in the way explained earlier, to be regarded as of outstanding importance in determining the contribution which plants make to the total environment, and then arranging the various groups recognized on this basis into a limited number of categories, adopting, because of its immediate practical utility the fiction that any single group can be made to fall, beyond dispute, into one or other of these categories.

This latter step, the chapter goes on to show, may conveniently be taken by first separating into a category of their own what may be called the isolated (or detached) groups of Dicotyledons, meaning by this those which have the most marked structural characteristics of their own and no very obvious linkings, through intermediates, with other groups. There is, of course, no final arbiter in recognizing these save personal opinion, which may not always be fully shared by others, but an impartial review of the groups so distinguished is likely to show that they share a good working measure of reality. These isolated groups number seventeen, and contain some 31,000 species, including the largest of all flowering plant families, the Compositae.

Next, there can be recognized in much the same way a category of what may be termed semi-isolated (or attached) groups. Many of these have, typically and generally, very distinctive characterziation, but they cannot be considered as truly isolated because they include minorities of members which are less specialized in form and which therefore link up with the more generalized Dicotyledons. The number of semi-isolated groups is reckoned at twenty-four, containing some 42,000 species, including such familiar plants as the Peas, the Labiates, the Crucifers and the Umbellifers.

The isolated and semi-isolated categories obviously contain the most easily definable constituents of the Dicotyledons and it is therefore not surprising that many of them are coterminous with taxonomic families.

In the central mass of Dicotyledons, analysis becomes increasingly difficult and resolves itself into separating this whole into a core, comprising those types (it is no longer possible to speak faithfully of groups) which show a minimum of distinctive characterization and separation, and into a peripheral part comprising the types which show some degree of specialization of one kind or another. These together complete the tally of the Dicotyledons, the former containing nearly 20,000 species and the latter probably upwards of 100,000.

Each of these great sections of the Dicotyledons has its own interest in relation to evolution but the peripheral part of the central mass especially so perhaps, for close examination shows that this can be pictured as made up of five sectors, each of which illustrates what may be reckoned as a major developmental theme. These are, respectively, a theme of *floral increase*, in which the flowers are larger than the norm and/or with a multiplication of parts, especially of the stamens; a theme of *floral decrease*, where an opposite state of affairs prevails; a theme of *floral zygomorphy*, in which one or other of various kinds of irregularity is found in the flowers; a theme of vegetative simplification expressed as the prevalence of the *herbaceous condition*; and a theme of *ecological specialization*.

A further result of the greater morphological (and consequently taxonomic) complexity of the Dicotyledons is that it is, in practice, impossible to represent them adequately in simple model form, or even by a wholly true-scale diagram. Nevertheless, a somewhat cruder sort of two-dimensional picture can be conveyed by imagining two simple concentric circles (core and periphery), the outer divided into five sectorially; a set of smaller and separate outer circles in contact with the periphery (the semi-isolated or attached groups); and an outermost, separated, set of circles (the isolated or detached groups). See figs. 97, 98.

The main object of Chapters 4, 5 and 6 was to display, as vividly as possible within the space available, both the grand evolutionary achievement of the Flowering Plants and the more salient of the innumerable facts which must be taken into account in any attempt to explain it. But these chapters also help us much better to appreciate many other more particular facets of the problem, and among these the division of the whole phylum into two great series of Monocotyledons and Dicotyledons is again one of the more conspicuous. In many directions there can be detected, without undue exercise of the imagination, constituents of the one which are, *in one respect or another*, the counterparts of constituents in the other. One of the most obvious is that of the Orchids and the Asclepiads, each of which is the sole pollinial group of its series; while both in the Pipeworts and in the Compositae there are capitulate pseudanthia, though expressed in different intensities. Floristically the Orchids and the Balsams are reminiscent of one another in general design of flower, which is, within each series, unique, and this shows how one and the same group may be the counterpart of more than one other according to the combination of characters it exhibits. Then there is the familiar example of the families Alismataceae and Ranunculaceae, so often canvassed in discussions about the origin of the Monocotyledons, but this is perhaps a little different because many believe that the actual relation between these is very close. Much the same is said sometimes of the Aroids and the Piperaceae, between which there are striking resemblances, but the Aroids have an even more notable parallelism, involving to some extent both function and form, with the Birthworts. A more general structural likeness is that between the families Dioscoreaceae and Menispermaceae, and, on a smaller taxonomic scale, that between the genera *Triglochin* and *Plantago* (see p. 343 also). There are also ecological parallels, ranging from such as that of the miniature water-lily-like genera *Hydrocleis* and *Limnanthemum*, through the general parallel between the freshwater members of the family Naiadaceae and the family Ceratophyllaceae, to the quite extraordinary thalloid counterparts, the Duckweeds and the Podostemads. But even more interesting are some of the subtler and less easily definable parallels, such as that of the Gingers and the Labiates; or that of the Spiderworts, in relation to other Monocotyledons, and the Melastomataceae in relation to other Dicotyledons. Even the wind-pollinated Palms, though so characteristic a Monocotyledon sort, are in some ways functionally the counterparts of the Catkin-bearers in the Dicotyledons.

These notable comparisons must not, however, obscure the fact that there are many unities in each series which have no obvious counterparts in the other, though it is more difficult to deny parallelism than to detect it. The Grasses, for instance, are such, for although there are sundry wind-pollinated herbaceous Dicotyledons there is no great coherent group of them. The Bamboos, again, cannot be compared with any of the larger Dicotyledons, while conversely the deliquescent and fully woody habit is virtually unexpressed in the Monocotyledons. There are also the two interesting points that the Monocotyledons, although they contain a number of saprophytes, have neither truly parasitic nor insectivorous members. On the other hand all the marine angiosperms belong to the Monocotyledons. In short the two great series show just such a combination of parallels and divergences as might be expected to result from a unified plan of evolution working upon diverse potentialities. The problem is what process of evolution and what kind of phylogenetic descent can have culminated in such an array of simultaneous similarities and differences?

The possibilities seem to be three. It may be that the counterparts and parallels are to be regarded as of more fundamental significance and that the two series exist because of some constantly repeated process of duplication—the production as it were, of Monocotyledon and Dicotyledon versions of almost every step or tendency. It may be that the two have been distinct ever since some single very remote phyletic dichotomy, and that they have subsequently repeatedly reflected one another's progress. Perhaps there has been some measure of both of these. Whichever, if any, may be true the problem is no less baffling. If the first is true then allowance must be made for a kind of evolutionary change or sequence which constantly repeats itself. If the second there must be visualized two quite distinct lines of descent in which, within the limits imposed by their essential differences, evolution has expressed itself in a whole number of corresponding conditions, a state of affairs which is difficult to explain unless, at least in many cases, the same evolutionary potentialities are to be found in both. If the third then both these explanations are involved. Whatever the truth it would seem to suggest unfamiliar aspects in the evolutionary story of these plants.

This analysis of the two series shows also how large a part of each consists of what are usually called "specialized" plants. Specialization in this context may be defined as the possession of peculiarity or uniqueness through what may, in evolutionary idiom, be spoken of as the accumulation of distinctiveness, a definition which rightly puts the chief emphasis on evolutionary divergence or departure. It is supposed, in short, that specialized plants and animals have gone off on evolutionary tacks of their own and have thereby acquired peculiar morphological personalities. This is a familiar biological conception and it has received considerable attention from evolutionists because it can sometimes (and especially in the higher animals) be associated directly and purposefully with mode of life, and thus with the two ideas of "adaptation" and "natural selection". At other times, however, specialization cannot be accounted for simply in these terms, and it is one of the most interesting examples of the distorting effects of unbalanced evolutionary study that the apparent sweet reason of the adaptive explanation in some animals so completely overshadows the many specializations in other animals and in plants which are clearly not susceptible to this explanation. The important element in the situation is, of course, the apparent purpose in most of the animal examples mentioned, and doubtless it is the absence of this in so many cases among plants that has led to their neglect, but whatever the cause the fact remains that specialization of this seemingly less purposeful kind is widespread in animals and plants: so much so, indeed, that it is scarcely an exaggeration to maintain that a satisfying elucidation of it would make a greater contribution to the better understanding of evolution than almost anything else.

It is here perhaps that the Flowering Plants are of greatest significance in the study of evolution, for not only is there remarkable specialization in many of them but it is pre-eminently of the sort that cannot be easily dismissed by the application of any principle of adaptation. At the same time, however, the very ubiquity of specialization in these plants makes for difficulty of description, for it is easy to miss the real points of issue in such a wide array of facts, and that is why the subject has been dealt with in two stages, the general and particular, the former comprising Chapters 4, 5 and 6.

CHAPTERS 7 AND 8 PROVIDE THE MORE PARTICULAR TREATMENT of specialization and may be considered together here. Almost any one of the larger isolated groups of Flowering Plants might have served as subject matter here, but there can hardly be a better choice than the Asclepiads, because these plants, chiefly on account of the details of their reproductive machinery, pose the kind of "adaptation" problem which has just been mentioned in an unusually inescapable way. Chapter 7 is a straightforward condensed account of these plants and need not be summarized here and Chapter 8 is an immediate consideration of some of the more important and interesting evolutionary points which this raises. To this extent the Asclepiads have therefore already been discussed and this part of the ground need not be re-trodden, but the present chapter affords a valuable opportunity of saying rather more about the over-riding interest of these plants, namely their significance in relation to the hypothesis of natural selection. Is it reasonable to believe that the peculiar features of these plants, and above all their pollination mechanism and their coronal elaboration, which are the bases of so much of their speciation, can have been produced by a process of natural selection or is it not?

If it be contended that natural selection offers an adequate explanation of the extraordinary state of affairs in this group, then it is necessary to explain exactly the nature of the mechanism by which it may have achieved its end, and, particularly, to point out the factors which, operating in this manner, may have decided and controlled the directions of evolutionary change. It is not enough merely to maintain the truth of natural selection in such circumstances as these; there must also be a competent idea of how it may have operated. If, on the other hand, natural selection is rejected as an explanation, it cannot merely be rejected in the case of the Asclepiads, it must also be rejected in respect of all those other examples which, though perhaps less striking, are so much of the same kind, and which provide, as earlier chapters clearly demonstrate, so much of the basis on which the taxonomy of the Flowering Plants is built. But if this explanation be denied so widely, is it possible to say, with any confidence, where it ever begins to be applicable? It is thus as a crucial test case in relation to the hypothesis of natural selection, rather than for any other reason, that the Asclepiads are given such prominence in this book. To them the reader is recommended to apply, not only all the familiar arguments in favour of the idea of natural selection, but also all those strangely less familiar arguments against it which occupy much space in the earliest literature of the subject and which have never to this day been satisfactorily countered.

CHAPTER 9 IS THE FIRST OF THREE having, as their subjects, still more particular evolutionary aspects of the Flowering Plants, and deals with the topics of floral aggregation and of the pseudanthium, both of which illustrate another pervasive problem which may be called that of *direction in evolution*, and which involves that familiar will-o'-the-wisp "primitiveness".

One of the most striking features of the organic world, in the sense of the part that it plays in making nature what it is, is the fact that the constituent characters of organisms are, by and large, severally variable, being expressed in different values in different individuals. For example every species of flowering plant has its own typical size of flower, but the actual dimensions are seldom precisely the same for any two or more individuals. The individuals of a species can therefore be arranged in keyboard fashion according to the size of their flowers. Similarly on a larger scale the keyboards of all species may be combined into a single super-keyboard representing the variation gamut of the character "flower size" throughout the Flowering Plants. In both cases the essence of this mode of depiction is the orderly arrangement of values, usually, but not necessarily, with the smaller at one end and the larger at the other.

What has just been described is of course an expression of living nature at any one time. It is a distribution of values in space only, and as such must long have been a familiar, if unformulated, character of the human environment. When, however, with the growth of the science of geology, knowledge of past floras and faunas increased, it soon became apparent that something like a keyboard of this sort would express much of variation in time. That is to say when fossils were arranged in chronological order they revealed in many respects a progressive change with time essentially of the same sort as the variation gamut in at least some of the characters of the present.

It is interesting to speculate on the relationship between these two expressions, but in practice it was the second of them, the recognition of variation or change in time, which made some theory of evolution almost inevitable, because the conclusion that the change itself was caused by the actual passage of one condition genealogically into another quickly became irresistible. Looked at logically the idea of an *evolutionary* process is, in the circumstances, a sound and justifiable working hypothesis to account for gradual chronological change, and allowed to stand by itself would, as such, be unassailable. There is, however, a temptation to apply the same reasoning more widely, so that there is current today the proposition that variation in space alone bears the same kind of relationship, while the opinion that variation between contemporary organisms is an expression of evolutionary procedure is now almost an axiom. Thus, ideas about variation in space and ideas about variation in time are now much in parallel, but there is a fundamental distinction between them which makes them, as working hypotheses, of very different values. This is the absence of a time sequence in matters concerning space alone.

The existence of a time sequence such as that in the fossil record is all important in evolutionary theory because, if rightly assessed, it is capable of telling us two things about the variation gamut, namely whether it has been continuous in direction, and which end of it came first. It is the absence of this orientation that makes the existence of variation gamuts in space alone so much less valuable as evidence of evolution in general, and of the courses it may have taken in particular. There is seldom anything in a contemporary or "living" variation gamut to determine whether, because the individuals can be arranged in a certain keyboard order, they are to be regarded as similarly related, or whether the sequence that the gamut provides is to be read in one direction or the other. This problem with its inherent intractability, has probably more bedevilled ideas about evolution than almost anything else, and Chapter 9 describes some striking examples of it.

But it is important to be clear on this matter. The comments which have been made are not intended to deny the possible truth of many evolutionary conclusions based on the study of contemporaneous variation, but merely to stress that they are often no more than intelligent guesses and, if true, no more than fortunate choices between two possibilities. That they indeed often amount to no more than this is shown by the ease with which it is possible to maintain diametrically opposite views with equal apparent logic and justification on so many of these questions.

On the other hand it is not entirely fair to regard these opinions as always without reason, for many of them are based on a series of propositions, often called the *criteria of primitiveness* which have been derived from a whole body of knowledge drawn from many sources, not least among them being the geological past. Unfortunately, it is tempting to transgress the proper limits of these conclusions, on the one hand by adding "criteria" which do not deserve the name because they have no historical sanctions, and on the other hand applying these and others without due regard for their appropriateness. This has a further and more subtle danger. In principle the stronger the evidence for any particular theoretical opinion, the more dispassionately may the facts be allowed to speak for themselves, and conversely, the less conclusive the evidence the greater is the temptation to use emotive words or ideas calculated to reinforce it, a proceeding referred to earlier as contributing much to the inexactness of some evolutionary literature.

The fact is that the sum of reliable knowledge about many of these things is less than it is comfortable to admit, and unless it is made the most of this fact may become embarrassingly apparent. There is also the point that unless assumptions of one kind or another are occasionally made, it becomes difficult to progress at all. We must always, therefore, be on the look-out for some *modus operandi* which, while preserving us from the dangers of the one will also save us from the excesses of the other, and this demands an austere attitude towards the drawing of conclusions from insufficient premisses. Chapter 9 illustrates this clearly, because, although there is a widespread opinion that strong floral aggregation is a derived condition, there is little or no real evidence that this is so. All that can strictly be said is that there is general throughout the Flowering Plants a very complete and wide variation gamut in respect of the aggregation of the flowers, and that this merges, in one direction, with what is an almost separate phenomenon, the pseudanthium. The facts of this matter present two evolutionary problems in particular, first that of the time sequence, if any, that the gamut reflects, that is to say whether or not there has been a chronological tendency towards aggregation, and second how and why the pseudanthial condition may have come about. The first has been sufficiently commented upon and the second is best dealt with in connection with the next chapter.

CHAPTER 10 MAY BE REGARDED AS A PROJECTION OF THE LAST because it is entirely devoted to the most remarkable expression of the pseudanthium, the Compositae. At the same time this is not the only evolutionary problem of these plants, and another will also be touched upon in due course.

How can the occurrence of pseudanthia be explained? In all the existing and antecedent circumstances (including the fact that pseudanthia are relatively uncommon) the most natural view is that they are specializations, being derived from some stock or stocks in the euanthial state, and on this view it is to be presumed that they are the end conditions of aggregation variation gamuts in time. But it is not quite so simple as this because the diagnostic feature of a pseudanthium is not mere aggregation but a degree of organization which results in the whole collection of flowers functioning, and looking, like one, and here there are two points to be noted. In the first place what can the initiating and implementing factors in this be, and in what manner can this organizational increment have become imported into the evolutionary trend? No simple intensification of aggregation or prolongation of that process alone will necessarily produce a true pseudanthium. There must arise at some point coincident and germane morphological features which will constitute this organization. To take a very simple case, it is the involucre of phyllaries which, more generally than anything else, makes a Composite capitulum what it is, but no such involucration is an *inescapable* result of ordinary floral aggregation, and there are many comparable degrees of aggregation which are without it. The Composite pseudanthium is essentially what it is because the aggregation has been accompanied by something more in the way of organization and co-ordination. The problem is to discover what kind of evolutionary process can have been responsible for this. To argue that at some point of time the development of an involucre becomes desirable or necessary still leaves unanswered the question of how this desirability or necessity becomes translated into its initiation.

Another possible opinion is that the pseudanthial condition is prior or primitive, and has itself given rise to the state of separate flowers. For just the reasons which make the first suggestion seem more plausible this is much less so, but the fact must be stressed that in terms of the real evidence available it is little, if at all, less probable. The only thing which could finally decide the issue is a knowledge of the past so complete as to reveal which condition came first. This is not forthcoming, and although such positive information as there is points, if only by default, to the first suggestion, it should be realized that this does not in fact preclude the possibility of the second. Moreover this second view has, in some ways, a good deal of common sense about it, because, if we can truly divest our minds of preconceived notions it is in some ways simpler to imagine evolution as working from small aggregated flowers towards the elaboration and individualism of larger separated flowers than the reverse.

A third possibility (see, especially, Wettstein, *loc. cit.* in Chapter 4 above) is that all bisexual and some unisexual flowers are in fact pseudanthia, each single stamen or group of stamens and each carpel or ovary representing unisexual flowers; the perianth, if present, presumably being of an involucral nature. At first this also sounds strange but again it must be remembered that there is no sufficient evidence which will automatically preclude it. It may be very unlikely to be true, but it is no more than this, and we must not lose sight of the difference between a high degree of improbability and certain impossibility.

With the two latter possibilities a further nice complication arises. If the pseudanthial condition is original, it would seem necessary to assume that those pseudanthia in which, today, the constituent flowers are fully organized and equipped, as is particularly the case with the Composites, must be more evolved and differentiated types than those in which, as in some Spurges, it is only with difficulty that the floral nature of the parts is revealed. It need only be observed that if this is indeed the case then the problem of the Composites becomes even more puzzling.

It seems then that there are two points to be remembered about the question of how the pseudanthium may have come about. One is that we are not truly in a position to preclude absolutely any one of three possibilities. The other is that if the more familiar explanation that the pseudanthium is an end product of aggregation is admitted, then we have to explain how those characters of the pseudanthium which are additional to mere aggregation may have come into being. Whatever view is taken there is no doubt that the pseudanthium is one of the most pregnant problems of evolution in the Flowering Plants.

The other Composite problem to be noticed here is not only different in kind, but does not seem to be presented by other pseudanthial plants, at least to the same degree. Its basis is the fact that although it is customary to think of the typical or "normal" Composite as something of the sunflower or daisy sort, this is not nearly so much the case as is generally imagined, the majority of the members of this group having something less than this completeness of design. Not only so but this relative incompleteness commonly takes the form of a marked diminution in the number of florets in the capitulum, even to the lowest limit of one, so that there are quite a number of cases in which, within the involucre of phyllaries delimiting the capitulum, there may be no more than one, and more often no more than five, florets. Moreover, in a significant total of instances such pauciflorous capitula are combined into capitula of a second order delimited by what are in fact super-involucres. Even more, there is, if description is correct, one case in which the process reaches the third degree. There are among Composites, then, and accompanied by a very complete series of intermediate conditions, three types of organization, namely the multiflorous and heteroflorous daisy type; the pauciflorous and homoflorous microcapitulate type; and the compound capitulate type. The question is what the evolutionary sequence and relations between these may be.

Here again there are the same sort of possibilities as before. Any one of the three may be original and from it the others may be descended, or more than one of them may have arisen quite independently. It is not proposed to plead the cause of any of these here, but to point out the evolutionary implications which each involves. Briefly, if the daisy type is prior then it must be supposed that evolution has, to a considerable extent, resulted, first in the gradual crystallization of the daisy type from non-capitulate stock (we cannot, on score of space alone deal with the possibility that the daisy type is the primitive flowering plant) and then in a resolution of this into something much more resembling an earlier state; this to be followed in certain instances by the restoration of the more specialized condition. Similarly, but conversely, if the one-flowered condition of the capitulum is prior, how can there be explained and whence may have come such things as the often typical involucre in such cases, and the usually exact structural resemblance between the single florets of these capitula and the individual florets of more multiflorous capitula? Again, if the compound capitulum is prior where has it come from and what were the stages in its evolution? Finally, if two or more of these types arose independently then it would seem that evolution has, from distinct sources, produced results of remarkable similarity.

The implications of the first, and most commonly held, of these possibilities are the most far-reaching because it pictures evolution as first culminating in the heteroflorous daisies; as giving rise, either concurrently or consecutively with this, to the more pauci-florous kinds of Composite; and, lastly, as having produced those members of the group which have compound capitula. It is in short difficult to avoid the conclusion that evolution has in some measure at least taken an oscillating course, and, if this is so, to avoid the more profound issue of what may be its evolutionary goal. This is the second special evolutionary problem of the Composites.

Many Composites bear surprisingly close resemblances to members of quite different families, sometimes in form, sometimes in function, sometimes in both. This similarity is often such in detail as to make almost inevitable the conclusion that there can be no really measurable difference of biological virtue between the two, or at least that none of the actual differences of design or behaviour amount to this. What, it may be asked, is the reward of evolution in these circumstances? If indeed it is correct that a member of the Compositae, which is usually regarded as one of the most specialized of Dicotyledonous families, is in practically the same biological state as a member of the Ranunculaceae, which is, as commonly, reckoned one of the least specialized groups, then the presumed greater amount of evolutionary change expressed by the structural differences between the two, seems meaningless. There is the further point that this great evolutionary change which produced the Compositae would seem most likely to have taken place while the Ranunculaceous stock continued relatively unchanged. This perplexing question can of course be begged by emphasizing how difficult it is to assess biological values, and in any particular case this may be a valid comment, but the phenomenon of which this is only an outstandingly clear instance is far too wide-spread among flowering plants to be discounted in this way. Indeed, it is not an overstatement that the whole display of evolutionary virtuosity in these plants indicates that a great deal of what is

called evolutionary change seems, in terms of the commoner
human conception of value, to be gratuitous.

CHAPTER 11 MAY, IN ITS TURN, BE RECKONED A PROJECTION
of its predecessor because its theme, superficial resemblance, is
only a more generalized expression of the same problem of the
evolutionary goal. For various reasons superficial resemblance is a
particularly troublesome subject. Its subjectiveness makes a neat
and orderly presentation difficult, and the very word superficial,
though it can hardly be replaced, has, through some of its other
associations, a slightly derogatory flavour. There is a traditional
contempt for resemblances of this kind which seems to be a relic
of the earlier days when so much biology was strict morphology
and so much morphology the search for homologies of phylogenetic
value. There is also to be remembered the confusion that has grown
up round the conception of "mimicry" and, to a lesser extent,
"convergence", both of which have their factual bases in likeness of
this sort.

About "mimicry" only one brief comment is called for here. It
is hard to find any serious general applications of this, so familiar in
some branches of biology, to plants, and it is difficult to resist the
conclusion that this absence of reference is a consequence of the
palpable obstacles which lie in the way of extending this conception to these organisms. This view, however, involves an *impasse*
which is not always realized. It may, indeed, well be that failure to
apply the idea of "mimicry" to plants is due to the manifest
absurdity of doing so, and this amounts to a tacit admission that it
is not properly so applicable, but the superficial resemblances on
which it rests are, as Chapter 11 amply shows, a common
feature of the Flowering Plants, just as they are of certain animal
groups. This being so it would seem clear that the real difficulty
lies in the circumstance that the conception of "mimicry" in
animals embodies an element of teleology which cannot easily be
extended to the plant world. If this is, indeed, so, and few biologists will disagree with the proposition, then the defence, so often
invoked when "mimicry" is called in question, that the term
means nothing more than mere likeness, must obviously fall.

Whatever the value of this comment on "mimicry" the broader
purpose of this chapter is to show how widespread among flowering

plants repetition of characters and, consequently, superficial resemblance, are, and how necessary it is to find a place for these within the body of evolutionary theory. This it does by presenting, within the framework of a simple classification, examples selected, among other reasons, for the frequency with which they can be paralleled by common observation. These need not be summarized again here but one or two points relating to them may be pursued a little further.

It must always be remembered that, because of the formal and functional differences between plants and animals, there is often a corresponding difference in the very nature of their morphological characters, and this bears closely on some aspects of superficial resemblance. Thus, among the higher animals the characters which, generally speaking, contribute most easily to superficial resemblance are those connected with the manner in which, or *how*, the animal nourishes itself—teeth, limbs, sense organs and the rest. On the other hand the characters which contribute most obviously to superficial resemblance in higher plants, such as the form of stems, leaves and inflorescences, have little direct bearing on the *method* of nutrition, though they may affect it in other ways. In plants it is the characters of the floral mechanism which are most clearly functional in this sense, determining as they do, the way in which pollination will be effected. Again, the kind of patterning which is so often the basis of superficial resemblance in insects, or the remarkable elaboration of the genitalia in some insect groups, are the result of characters which have little or no direct connection with the nature or source of the animals' food. On this basis it would seem that a distinction must be drawn between characteristics of what may be called behaviour as illustrated for instance by the body shapes of mammals and by the floral mechanisms of plants, and characters more particularly of appearance only, such as the more vegetative features of plants and the patterning of insects. All these may contribute to superficial resemblance in various ways and with varying significance, but it certainly seems that here again, as in certain other respects, plants and animals are the converse, or at any rate the complement, of one another.

Although this idea may be something of an over-simplification, and although there may be many characters which do not readily conform to either definition, it can be of help in the study of evolution, if only as a constant reminder that the characters which make up an organism *in toto* are not all likely to be strictly comparable and equatable. Especially is it worth noting that while it is the change in the more "functional" characters which provides so much of the element of novelty that is the whole essence of progressive evolutionary development, it is change in the more "formal" characters which provides so much of the material used in the study of variation. This point of view also helps to put the present position with regard to evolutionary knowledge clearly by saying that while, thanks to the development of genetics in particular, a great deal is now known about the nature, origin, behaviour and fate of these more formal characters, comparatively little is yet known about the factors controlling the origin of novelties.

To admit the superficial resemblance between certain plants is not always also fully to appreciate its implications. The structural and visual "drawing together" of the two plants may, in theory at least, come about either by a similar departure from some norm on the part of each, or by departure from that norm largely or entirely in one plant only. In other words assuming that progressive change is involved, this may be shared by both or restricted partly or wholly to one or other. There are various reasons why the former or "balanced" state is harder to detect but the fact seems to remain that in a notable proportion of cases the resemblance is "unbalanced", that is to say is due to a much more considerable departure from the usual in one of the plants than in the other. Even in the remarkable cases in which Composites are concerned the superficial resemblance is nearly always due to the fact that the member of that family lacks, at least in some measure, the features which would normally quickly betray its true affinity to the eye. The interesting point is that this lack does not result in a visually unique plant but in one more like some other plant.

To sum up, Chapter 11 shows that the kind of similarity which has been called superficial resemblance is neither a thing of rarity nor a figment of the imagination, but is to be reckoned a not uncommon feature of the Flowering Plants. It is this frequency of occurrence which is its most important aspect here for it points clearly to the fact that again and again in the course of their evolution there has appeared, in some or other lineage, a condition more or less foreign to it and more characteristic of others. That is to say various detailed and particular character values have appeared more than once, in different lineages, and in varying degrees independently of one another. These characters have, indeed, been repeated, and what is said in Chapter 11 demonstrates that evolutionary *repetition* of this sort has been a feature in the history of these plants. Moreover the situation suggests that the differences between the "families" of flowering plants are not so much inherent differences of kind as differences of character combination, both qualitative and quantitative. Each seems to have its own basic combination of functional character values, and the more stable this is the more easily recognizable and "natural" the family is. Looked at from this point of view superficial resemblance is seen to be due, essentially, to the presence in one plant of a character or characters equally or even more commonly associated with some other kind of plant, and in this conception the principle of repetition is integral.

What the true significance of this repetition may be is by no means apparent, but there are various possibilities which cannot be totally ignored here even though the hope of demonstrating any one of them in the present state of our knowledge is so slight that they can really only be expressed in the form of questions. Does repetition of this sort mean that there is, effectively, some limitations in the possibilities of evolution? Are these resemblances simply the inevitable ultimate reappearances of rare combinations among a comparatively small number of ingredients? Or are they conditions which, within an almost or quite unlimited range of possibility, appear with more than average and expected frequency? Each of these questions invites still further speculation. Real limitation of possibility, in the sense that the number of character factors in evolution is to be regarded as strictly finite, would account for some of the features of evolution in general, such as the phenomenon of repetition, but it would not account for others, and especially would it not account for the large-scale march of evolution, which has been so much a matter of the development of novelty. On the other hand if there is no real limitation of possibility and if every condition may, in theory at least, be attainable, then an explanation must be sought of why certain values and characters should so commonly reappear in the way which has been called repetition. Can it be that the occurrence of certain values or characters has a favourable predisposing effect on the development of others? Or can it be that the occurrence of certain values and characters is prejudicial or fatal to the development of others? Again, is it to be imagined that gradually changing values and characters tend, that is to say gravitate, more easily towards some few particular expressions rather than others? And, finally, one of the most interesting speculations of all, have values and characters anything like a natural and inevitable sequence, in virtue of which once a given state is achieved some future state becomes predestined? Such at least are some of the thoughts which a study of the Flowering Plants in general, and of superficial resemblance in particular, must bring to mind. Admittedly it is easy enough to pose such a string of questions as this, and there may be but little immediate profit in doing so, but this does at any rate show how few are those to which a really satisfying answer, has, so far, been found.

FINALLY LET US EXAMINE, with a wider appreciation, the picture which this book has outlined. It is a picture of a vast multitude of differing plants—plants of almost every size and shape within the definition of their phylum; plants occupying, in every degree of rarity or plenty, almost all the kinds of situation that land provides; plants so multifarious that the sorts distinguishable by the human senses alone amount, even in the simplified terms of taxonomy, to nearly a quarter of a million, a number in itself so great that there is scarcely a spot on the surface of the earth that does not harbour scores of them. Can we explain how this astonishing pageant of life has come into being?

That it has been built up through the ages by some great continuing process of elaborative morphological change, or evolution, is the most acceptable suggestion that has so far been made, and accordingly biologists are, by and large, evolutionists today, but it is worth while to pause for a moment and to consider what it is that is really implied by this opinion. Our present knowledge of biology, reinforced by traditional scientific teaching, encourages us to suppose that this continuity of change which constitutes evolution has been complete, not only in the sense that there has been no break in the great chain of generations, but also in the sense that offspring have, throughout it, varied from their parents only within the dimensions of our present every-day experience of heredity. The first of these two beliefs is unequivocal enough, and, in its most absolute form, in which a single origin of life on the earth is assumed, it means that every individual living organism existing today has the same unbroken continuity of descent from some unimaginably remote primordial ancestor. Even allowing for the contingency that life may, in the distant past, have originated more than once, it means very little less than this. The second belief is much less simple because it rests on the proposition that what is observed to be true today of the relation between parent and offspring has invariably and always been true, and this may or may not be so. Yet it is only in combination that these two beliefs constitute evolution as it is usually pictured today and the crucial question is therefore whether *both* can be accepted *in terms of one another*. Can we believe that there has been complete genealogical continuity *on that scale of intergenerational change which is familiar today* or must we suppose that on one or more occasions in the course of geological time, this very gradual evolution has been accelerated or short-circuited by some process of a different value, and even perhaps of a different nature? Our decision must of course depend on the worth of the evidence available and when this is considered the result is interesting. On this score of evidence it is the second belief, namely that concerned with the dimensions of intergenerational change, which appears to be the less equivocal, for although it is inductive it is at least based on some measure of direct observation and experience. On the contrary the first belief (that there has been complete genealogical continuity) involves the inherent impossibility of recapturing the past and the consequent necessity of relying on circumstantial evidence, the strength of which cannot be so satisfactorily assessed. In this particular proposition the circumstantial evidence is that afforded by the fossil record and this indeed shows beyond reasonable doubt that the earth has, in the course of geological time, been populated by a succession of life forms each, on the whole, more elaborate than the one before it. It not unnaturally seems at first sight therefore that the two beliefs are complementary propositions which together make up a complete and adequate theoretical whole, but further consideration reveals that this is not so and that there is a weak link in the argument. This flaw is the fact that the fossil record, though disclosing an orderly succession of biological types in time, does not, and it would seem cannot, reveal whether the members of this succession constitute an unbroken genealogy in which each has been the immediate progeny of that which preceded it. This is the issue which the record of the rocks does not clinch for us. Consequently there is no direct evidence as to whether this grand procession of nature in the past has come about wholly, partly, or not at all by the gradual change of organisms into one another which, under the name of transformism, is the essential part of the concept of *organic evolution* as distinct from other sequences to which the word evolution is often loosely attached, such as the "evolution" of machines or of ideas.

It may be contended that this is an ungenerous appreciation of the fossil record and that the evidence for transformism in it is much stronger than has been suggested, but it is difficult to accept this estimate of the situation. It is true that some phases of the record show, especially along certain lines, successions of forms so closely knit as to make it hard to resist the conclusion that they do indeed represent direct series of transformations in which each has given rise to its immediate successor in time, but it is also true, and in a much more real way, that the record not only has gaps in many places, but that these are often just where the absence of reliable positive information is most frustrating and disturbing.

Recapitulating, although there may be many germane, as well as more collateral, reasons for believing that the biological past has been a gradual transformism of the kind depicted in current theories of evolution, there is no decisive evidence that this has been the case and thus there are no proper grounds for excluding from free consideration the possibility that the facts may be, at least in part, susceptible to some other explanation. That there has been progressive change with time seems to admit of no argument, but what kind of change may have been most largely involved and how far that change has been consistent in rate and sort are questions that still remain unanswered.

However distasteful this conclusion may be it is a fair and correct assessment of the situation regarding evolutionary theories today and we must therefore enquire here what light, if any, a study of the

Flowering Plants throws upon it. For one paramount reason, namely the absence of any effective palaeontological history of these plants, the answer can only be that their *direct* influence on this problem is comparatively small, but more *indirectly* many of their features, and especially their huge numbers, give them considerable significance. If a new biological type is not the result of transformism it must, as far as we are capable of judging, arise in some way within the meaning of the words *de novo*, and that is something quite beyond our experience. That we have no such experience, however, is not in itself a reason for excluding the possibility of an origin of this sort in all cases, though it does afford reasonable grounds for doubting whether this has ever been a general and wide-spread phenomenon, because if this had been the case the less likely that we should continue to be without any evidence of it. On this argument then the very numbers of the Flowering Plants tend to support the opinion that they are, at least in all essentials, the result of biological transformism. At the same time we cannot totally ignore the possibility that occasionally in time and space, if no more frequently, there may have been involved an event of a different kind.

Similarly but conversely, it is their great numbers that make it difficult to accept, without some hesitation, the view that the Flowering Plants are wholly the result of transformism in its stricter meaning. The question is whether we can really believe the past to have been so long and the necessary impetus to change so sustained that each member of this extraordinarily varied section of the living world is sib to every other and owes its existence to a process of gradual structural divergence. Especially is this difficult if the Flowering Plants are looked upon as monophyletic, but even if they are regarded as to some degree polyphyletic the problem is not much lessened, for this opinion imports fresh complexities into it.

It is from such doubts as these about transformism that much of the lack of unanimity of opinion in matters evolutionary arises, and it is notably with reference to this problem that many evolutionary hypotheses differ from one another. This is perhaps best illustrated by referring to three of the more familiar philosophies which have grown up within the evolutionary conception, to each of which a consideration of the Flowering Plants is very relevant. Of these, one has it that evolution is the slowly cumulative effect, over immense periods of time, of some inherent instability and inevitability of change which is part and parcel of the grand design of nature, and which is more or less independent of direct environmental influences. The second has it that evolution has come about as a result of a never-ending struggle for existence in which survival goes to those possessing characteristics of greatest influence in this struggle, and in which the weaker in this respect go to the wall. The third has it that evolution has consisted largely or entirely of a long series of comparatively sudden and considerable changes or mutations, brought about, it may be, by the immediate influences of the environment. To mention these three in this brief way is not to suggest that they collectively possess a monopoly of the truth; or that they are mutually exclusive of one another; or even indeed that they are directly and properly comparable. They do, however, express three of the most popular hypotheses of evolution, for the first embodies the general concept underlying *theories of orthogenesis*; the second comprehends the essential principle of the *theory of natural selection*; and the third is the basis of what may comprehensively be called the *mutation theory*. The bearing of these hypotheses on the problem of transformism is that each of them amounts to, or offers, an attempt to mitigate its intractability. Orthogenesis does this by predicating an inherent impulse towards change which makes it unnecessary to seek any leading external cause for this. Natural selection, on the other hand, predicates an external explanation of continued change which is severely empiric. Mutation predicates that change may, at least on occasion, be saltatory and thus potentially capable of reducing the time requirement to almost any extent.

Tested against the factual picture of the Flowering Plants these theories fare very differently. Regarding orthogenesis it can be said, though perhaps only because of the very nature of the proposition, that there is no major aspect of these plants which is patently inconsistent with it, nor any important direction in which they cannot be accommodated within its bounds. As regards natural selection it can only be said that it is difficult to avoid the conclusion that the circumstances exhibited by the Flowering Plants, and especially such of them as have been noticed in this book, are in many important respects inexplicable on this theory. This is not to deny that there must doubtless have been, and continue to be, times when, in nature, one individual or group of individuals gives way before another, but that such a vast array of life as these plants show can ever have come about by a process which, by its very nature, must be one of elimination, or can have maintained itself, with its almost infinite gradation of form, numbers and environment, in the face of some mortal struggle, is very hard to believe. Nor is it easy to see how those innumerable natural communities which are the foundations of ecology, and to which the Flowering Plants contribute so much, can ever have developed unless the whole scheme of nature has been one, not of antagonism, but of expanding opportunity and mutual contribution. As regards mutation it can be said not only that there is no major aspect of the Flowering Plants which is at variance with it, but that many of the facts are peculiarly appropriate to an explanation of this kind. But this is not all. Mutation theories lack precision in that they do not always define or estimate quantitatively the magnitude of the changes they postulate so that while, in one direction, there is no theoretical limit to the size of mutations, it is possible in the other direction to call quite small changes by this name. It is the particular significance of the Flowering Plants in this connection that much about them suggests the wide occurrence of large-scale mutation.

The conclusions with regard to the bearing of the Flowering Plants on these various hypotheses are, therefore, that these plants, though providing little positive evidence in favour of orthogenesis, show even fewer indications of evidence against it; that they reveal no evidence that natural selection has played an important part in their evolution, though they provide many indications that this has not, in fact, been the case; and that they suggest that mutation, on a comparatively large scale, has been a frequent and potent factor in their history.

Since this chapter has been, in effect, a *résumé* of those preceding it, it has been called a summary, but even so it is far too long to leave as it is, and the final step must now be taken of condensing into the briefest possible terms what seem, to the writer, to be the conclusions to which this study of the Flowering Plants has led. These are:

1. If we are right in thinking evolution to have been the cause of the present multifariousness of organisms, that evolution must, in view of the essential unity of the life expression, be regarded as having been fundamentally the same process in plants and animals.

2. Although there is this underlying unity, the details of evolution, and especially their respective emphases, must because of the basic ecological relationship between the plant and animal kingdoms, have been very different in the two. In particular the evolution of green plants may be regarded, because of their autotrophism, as having been independent, while the evolution of animals must, because of their immediate dependence on the plant world, have been conditioned to some extent by that of plants. This difference may be expressed by using the word *consequent* (following as a natural sequence of mundane environmental conditions) to describe the evolution of plants, and the words *subsequent* (following in time) and *obsequent* (compliant with or obedient to (that of plants)) to describe the evolution of animals.

3. To a very considerable degree evolution in the Flowering Plants appears, when subjected to human estimation, to have been gratuitous or motiveless, using these words to mean that it has been change because change is the innate order of nature rather than change because change confers biological superiority. It is not easy to detect, in the contemporary differences between flowering plants which are the expression of these changes, many which can claim to have selection value, or which amount to adaptation in the strict Darwinian sense of that word.

4. An examination of its achievement in the Flowering Plants leaves an abiding impression that the evolution of these plants has been, in essence, a process of continuing automatic impetus, and the expression of a peculiar dynamism of nature which is the antithesis of stagnation.

5. The study, in particular, of floral form in the Flowering Plants indicates that their evolution has embodied an important element of repetition, which notably takes the two forms of the repeated expression of the same kind of change in the same lineage and of the achievement of similar effects in different and unrelated lineages.

6. The impression that the Flowering Plants give of gratuitous and repetitive evolution suggests that much of their evolutionary

pattern is elaborative and well represented by the analogy of the kaleidoscope,* and that this part of the pattern accounts for most of their multifariousness. The other part comprises the more inherent and coherent sequence which constitutes the grand progressive march of evolution. This in turn suggests the existence of two coincident kinds of evolutionary change, one of which may be compared with the movement of a stream itself, and the other with the ripples on its surface.

7. Little or nothing in this picture of evolution in the Flowering Plants supports the view that they are the product of any highly competitive and eliminative plan of nature. On the contrary, it suggests that nature is expansive rather than contractile and that no matter what new characters or new combinations of old characters change with time may present they are all able to find an existence somewhere in the scheme of things.

* *Cf.* Willis, J. C. *The Birth and Spread of Plants*, Geneva, 1949.

INDEX

A CATALOGUE OF SELECTED DOVER BOOKS
IN ALL FIELDS OF INTEREST

A CATALOGUE OF SELECTED DOVER BOOKS
IN ALL FIELDS OF INTEREST

AMERICA'S OLD MASTERS, James T. Flexner. Four men emerged unexpectedly from provincial 18th century America to leadership in European art: Benjamin West, J. S. Copley, C. R. Peale, Gilbert Stuart. Brilliant coverage of lives and contributions. Revised, 1967 edition. 69 plates. 365pp. of text.

21806-6 Paperbound $3.00

FIRST FLOWERS OF OUR WILDERNESS: AMERICAN PAINTING, THE COLONIAL PERIOD, James T. Flexner. Painters, and regional painting traditions from earliest Colonial times up to the emergence of Copley, West and Peale Sr., Foster, Gustavus Hesselius, Feke, John Smibert and many anonymous painters in the primitive manner. Engaging presentation, with 162 illustrations. xxii + 368pp.

22180-6 Paperbound $3.50

THE LIGHT OF DISTANT SKIES: AMERICAN PAINTING, 1760-1835, James T. Flexner. The great generation of early American painters goes to Europe to learn and to teach: West, Copley, Gilbert Stuart and others. Allston, Trumbull, Morse; also contemporary American painters—primitives, derivatives, academics—who remained in America. 102 illustrations. xiii + 306pp.

22179-2 Paperbound $3.50

A HISTORY OF THE RISE AND PROGRESS OF THE ARTS OF DESIGN IN THE UNITED STATES, William Dunlap. Much the richest mine of information on early American painters, sculptors, architects, engravers, miniaturists, etc. The only source of information for scores of artists, the major primary source for many others. Unabridged reprint of rare original 1834 edition, with new introduction by James T. Flexner, and 394 new illustrations. Edited by Rita Weiss. 6⅝ x 9⅝.

21695-0, 21696-9, 21697-7 Three volumes, Paperbound $15.00

EPOCHS OF CHINESE AND JAPANESE ART, Ernest F. Fenollosa. From primitive Chinese art to the 20th century, thorough history, explanation of every important art period and form, including Japanese woodcuts; main stress on China and Japan, but Tibet, Korea also included. Still unexcelled for its detailed, rich coverage of cultural background, aesthetic elements, diffusion studies, particularly of the historical period. 2nd, 1913 edition. 242 illustrations. lii + 439pp. of text.

20364-6, 20365-4 Two volumes, Paperbound $6.00

THE GENTLE ART OF MAKING ENEMIES, James A. M. Whistler. Greatest wit of his day deflates Oscar Wilde, Ruskin, Swinburne; strikes back at inane critics, exhibitions, art journalism; aesthetics of impressionist revolution in most striking form. Highly readable classic by great painter. Reproduction of edition designed by Whistler. Introduction by Alfred Werner. xxxvi + 334pp.

21875-9 Paperbound $3.00

VISUAL ILLUSIONS: THEIR CAUSES, CHARACTERISTICS, AND APPLICATIONS, Matthew Luckiesh. Thorough description and discussion of optical illusion, geometric and perspective, particularly; size and shape distortions, illusions of color, of motion; natural illusions; use of illusion in art and magic, industry, etc. Most useful today with op art, also for classical art. Scores of effects illustrated. Introduction by William H. Ittleson. 100 illustrations. xxi + 252pp.

21530-X Paperbound $2.00

A HANDBOOK OF ANATOMY FOR ART STUDENTS, Arthur Thomson. Thorough, virtually exhaustive coverage of skeletal structure, musculature, etc. Full text, supplemented by anatomical diagrams and drawings and by photographs of undraped figures. Unique in its comparison of male and female forms, pointing out differences of contour, texture, form. 211 figures, 40 drawings, 86 photographs. xx + 459pp. 5⅜ x 8⅜.

21163-0 Paperbound $3.50

150 MASTERPIECES OF DRAWING, Selected by Anthony Toney. Full page reproductions of drawings from the early 16th to the end of the 18th century, all beautifully reproduced: Rembrandt, Michelangelo, Dürer, Fragonard, Urs, Graf, Wouwerman, many others. First-rate browsing book, model book for artists. xviii + 150pp. 8⅜ x 11¼.

21032-4 Paperbound $2.50

THE LATER WORK OF AUBREY BEARDSLEY, Aubrey Beardsley. Exotic, erotic, ironic masterpieces in full maturity: Comedy Ballet, Venus and Tannhauser, Pierrot, Lysistrata, Rape of the Lock, Savoy material, Ali Baba, Volpone, etc. This material revolutionized the art world, and is still powerful, fresh, brilliant. With *The Early Work*, all Beardsley's finest work. 174 plates, 2 in color. xiv + 176pp. 8⅛ x 11.

21817-1 Paperbound $3.00

DRAWINGS OF REMBRANDT, Rembrandt van Rijn. Complete reproduction of fabulously rare edition by Lippmann and Hofstede de Groot, completely reedited, updated, improved by Prof. Seymour Slive, Fogg Museum. Portraits, Biblical sketches, landscapes, Oriental types, nudes, episodes from classical mythology—All Rembrandt's fertile genius. Also selection of drawings by his pupils and followers. "Stunning volumes," *Saturday Review*. 550 illustrations. lxxviii + 552pp. 9⅛ x 12¼.

21485-0, 21486-9 Two volumes, Paperbound $10.00

THE DISASTERS OF WAR, Francisco Goya. One of the masterpieces of Western civilization—83 etchings that record Goya's shattering, bitter reaction to the Napoleonic war that swept through Spain after the insurrection of 1808 and to war in general. Reprint of the first edition, with three additional plates from Boston's Museum of Fine Arts. All plates facsimile size. Introduction by Philip Hofer, Fogg Museum. v + 97pp. 9⅜ x 8¼.

21872-4 Paperbound $2.00

GRAPHIC WORKS OF ODILON REDON. Largest collection of Redon's graphic works ever assembled: 172 lithographs, 28 etchings and engravings, 9 drawings. These include some of his most famous works. All the plates from *Odilon Redon: oeuvre graphique complet*, plus additional plates. New introduction and caption translations by Alfred Werner. 209 illustrations. xxvii + 209pp. 9⅛ x 12¼.

21966-8 Paperbound $4.50

DESIGN BY ACCIDENT; A BOOK OF "ACCIDENTAL EFFECTS" FOR ARTISTS AND DESIGNERS, James F. O'Brien. Create your own unique, striking, imaginative effects by "controlled accident" interaction of materials: paints and lacquers, oil and water based paints, splatter, crackling materials, shatter, similar items. Everything you do will be different; first book on this limitless art, so useful to both fine artist and commercial artist. Full instructions. 192 plates showing "accidents," 8 in color. viii + 215pp. 8⅜ x 11¼. 21942-9 Paperbound $3.75

THE BOOK OF SIGNS, Rudolf Koch. Famed German type designer draws 493 beautiful symbols: religious, mystical, alchemical, imperial, property marks, runes, etc. Remarkable fusion of traditional and modern. Good for suggestions of timelessness, smartness, modernity. Text. vi + 104pp. 6⅛ x 9¼. 20162-7 Paperbound $1.25

HISTORY OF INDIAN AND INDONESIAN ART, Ananda K. Coomaraswamy. An unabridged republication of one of the finest books by a great scholar in Eastern art. Rich in descriptive material, history, social backgrounds; Sunga reliefs, Rajput paintings, Gupta temples, Burmese frescoes, textiles, jewelry, sculpture, etc. 400 photos. viii + 423pp. 6⅜ x 9¾. 21436-2 Paperbound $5.00

PRIMITIVE ART, Franz Boas. America's foremost anthropologist surveys textiles, ceramics, woodcarving, basketry, metalwork, etc.; patterns, technology, creation of symbols, style origins. All areas of world, but very full on Northwest Coast Indians. More than 350 illustrations of baskets, boxes, totem poles, weapons, etc. 378 pp. 20025-6 Paperbound $3.00

THE GENTLEMAN AND CABINET MAKER'S DIRECTOR, Thomas Chippendale. Full reprint (third edition, 1762) of most influential furniture book of all time, by master cabinetmaker. 200 plates, illustrating chairs, sofas, mirrors, tables, cabinets, plus 24 photographs of surviving pieces. Biographical introduction by N. Bienenstock. vi + 249pp. 9⅞ x 12¾. 21601-2 Paperbound $4.00

AMERICAN ANTIQUE FURNITURE, Edgar G. Miller, Jr. The basic coverage of all American furniture before 1840. Individual chapters cover type of furniture—clocks, tables, sideboards, etc.—chronologically, with inexhaustible wealth of data. More than 2100 photographs, all identified, commented on. Essential to all early American collectors. Introduction by H. E. Keyes. vi + 1106pp. 7⅞ x 10¾. 21599-7, 21600-4 Two volumes, Paperbound $11.00

PENNSYLVANIA DUTCH AMERICAN FOLK ART, Henry J. Kauffman. 279 photos, 28 drawings of tulipware, Fraktur script, painted tinware, toys, flowered furniture, quilts, samplers, hex signs, house interiors, etc. Full descriptive text. Excellent for tourist, rewarding for designer, collector. Map. 146pp. 7⅞ x 10¾. 21205-X Paperbound $2.50

EARLY NEW ENGLAND GRAVESTONE RUBBINGS, Edmund V. Gillon, Jr. 43 photographs, 226 carefully reproduced rubbings show heavily symbolic, sometimes macabre early gravestones, up to early 19th century. Remarkable early American primitive art, occasionally strikingly beautiful; always powerful. Text. xxvi + 207pp. 8⅜ x 11¼. 21380-3 Paperbound $3.50

ALPHABETS AND ORNAMENTS, Ernst Lehner. Well-known pictorial source for decorative alphabets, script examples, cartouches, frames, decorative title pages, calligraphic initials, borders, similar material. 14th to 19th century, mostly European. Useful in almost any graphic arts designing, varied styles. 750 illustrations. 256pp. 7 x 10. 21905-4 Paperbound $4.00

PAINTING: A CREATIVE APPROACH, Norman Colquhoun. For the beginner simple guide provides an instructive approach to painting: major stumbling blocks for beginner; overcoming them, technical points; paints and pigments; oil painting; watercolor and other media and color. New section on "plastic" paints. Glossary. Formerly *Paint Your Own Pictures.* 221pp. 22000-1 Paperbound $1.75

THE ENJOYMENT AND USE OF COLOR, Walter Sargent. Explanation of the relations between colors themselves and between colors in nature and art, including hundreds of little-known facts about color values, intensities, effects of high and low illumination, complementary colors. Many practical hints for painters, references to great masters. 7 color plates, 29 illustrations. x + 274pp.
20944-X Paperbound $2.75

THE NOTEBOOKS OF LEONARDO DA VINCI, compiled and edited by Jean Paul Richter. 1566 extracts from original manuscripts reveal the full range of Leonardo's versatile genius: all his writings on painting, sculpture, architecture, anatomy, astronomy, geography, topography, physiology, mining, music, etc., in both Italian and English, with 186 plates of manuscript pages and more than 500 additional drawings. Includes studies for the Last Supper, the lost Sforza monument, and other works. Total of xlvii + 866pp. 7⅞ x 10¾.
22572-0, 22573-9 Two volumes, Paperbound $11.00

MONTGOMERY WARD CATALOGUE OF 1895. Tea gowns, yards of flannel and pillow-case lace, stereoscopes, books of gospel hymns, the New Improved Singer Sewing Machine, side saddles, milk skimmers, straight-edged razors, high-button shoes, spittoons, and on and on . . . listing some 25,000 items, practically all illustrated. Essential to the shoppers of the 1890's, it is our truest record of the spirit of the period. Unaltered reprint of Issue No. 57, Spring and Summer 1895. Introduction by Boris Emmet. Innumerable illustrations. xiii + 624pp. 8½ x 11⅝.
22377-9 Paperbound $6.95

THE CRYSTAL PALACE EXHIBITION ILLUSTRATED CATALOGUE (LONDON, 1851). One of the wonders of the modern world—the Crystal Palace Exhibition in which all the nations of the civilized world exhibited their achievements in the arts and sciences—presented in an equally important illustrated catalogue. More than 1700 items pictured with accompanying text—ceramics, textiles, cast-iron work, carpets, pianos, sleds, razors, wall-papers, billiard tables, beehives, silverware and hundreds of other artifacts—represent the focal point of Victorian culture in the Western World. Probably the largest collection of Victorian decorative art ever assembled—indispensable for antiquarians and designers. Unabridged republication of the Art-Journal Catalogue of the Great Exhibition of 1851, with all terminal essays. New introduction by John Gloag, F.S.A. xxxiv + 426pp. 9 x 12.
22503-8 Paperbound $5.00

A History of Costume, Carl Köhler. Definitive history, based on surviving pieces of clothing primarily, and paintings, statues, etc. secondarily. Highly readable text, supplemented by 594 illustrations of costumes of the ancient Mediterranean peoples, Greece and Rome, the Teutonic prehistoric period; costumes of the Middle Ages, Renaissance, Baroque, 18th and 19th centuries. Clear, measured patterns are provided for many clothing articles. Approach is practical throughout. Enlarged by Emma von Sichart. 464pp. 21030-8 Paperbound $3.50.

Oriental Rugs, Antique and Modern, Walter A. Hawley. A complete and authoritative treatise on the Oriental rug—where they are made, by whom and how, designs and symbols, characteristics in detail of the six major groups, how to distinguish them and how to buy them. Detailed technical data is provided on periods, weaves, warps, wefts, textures, sides, ends and knots, although no technical background is required for an understanding. 11 color plates, 80 halftones, 4 maps. vi + 320pp. 6⅛ x 9⅛. 22366-3 Paperbound $5.00

Ten Books on Architecture, Vitruvius. By any standards the most important book on architecture ever written. Early Roman discussion of aesthetics of building, construction methods, orders, sites, and every other aspect of architecture has inspired, instructed architecture for about 2,000 years. Stands behind Palladio, Michelangelo, Bramante, Wren, countless others. Definitive Morris H. Morgan translation. 68 illustrations. xii + 331pp. 20645-9 Paperbound $3.00

The Four Books of Architecture, Andrea Palladio. Translated into every major Western European language in the two centuries following its publication in 1570, this has been one of the most influential books in the history of architecture. Complete reprint of the 1738 Isaac Ware edition. New introduction by Adolf Placzek, Columbia Univ. 216 plates. xxii + 110pp. of text. 9½ x 12¾. 21308-0 Clothbound $12.50

Sticks and Stones: A Study of American Architecture and Civilization, Lewis Mumford.One of the great classics of American cultural history. American architecture from the medieval-inspired earliest forms to the early 20th century; evolution of structure and style, and reciprocal influences on environment. 21 photographic illustrations. 238pp. 20202-X Paperbound $2.00

The American Builder's Companion, Asher Benjamin. The most widely used early 19th century architectural style and source book, for colonial up into Greek Revival periods. Extensive development of geometry of carpentering, construction of sashes, frames, doors, stairs; plans and elevations of domestic and other buildings. Hundreds of thousands of houses were built according to this book, now invaluable to historians, architects, restorers, etc. 1827 edition. 59 plates. 114pp. 7⅞ x 10¾. 22236-5 Paperbound $3.50

Dutch Houses in the Hudson Valley Before 1776, Helen Wilkinson Reynolds. The standard survey of the Dutch colonial house and outbuildings, with constructional features, decoration, and local history associated with individual homesteads. Introduction by Franklin D. Roosevelt. Map. 150 illustrations. 469pp. 6⅝ x 9¼. 21469-9 Paperbound $5.00

THE ARCHITECTURE OF COUNTRY HOUSES, Andrew J. Downing. Together with Vaux's *Villas and Cottages* this is the basic book for Hudson River Gothic architecture of the middle Victorian period. Full, sound discussions of general aspects of housing, architecture, style, decoration, furnishing, together with scores of detailed house plans, illustrations of specific buildings, accompanied by full text. Perhaps the most influential single American architectural book. 1850 edition. Introduction by J. Stewart Johnson. 321 figures, 34 architectural designs. xvi + 560pp.
22003-6 Paperbound $4.00

LOST EXAMPLES OF COLONIAL ARCHITECTURE, John Mead Howells. Full-page photographs of buildings that have disappeared or been so altered as to be denatured, including many designed by major early American architects. 245 plates. xvii + 248pp. 7⅞ x 10¾. 21143-6 Paperbound $3.50

DOMESTIC ARCHITECTURE OF THE AMERICAN COLONIES AND OF THE EARLY REPUBLIC, Fiske Kimball. Foremost architect and restorer of Williamsburg and Monticello covers nearly 200 homes between 1620-1825. Architectural details, construction, style features, special fixtures, floor plans, etc. Generally considered finest work in its area. 219 illustrations of houses, doorways, windows, capital mantels. xx + 314pp. 7⅞ x 10¾. 21743-4 Paperbound $4.00

EARLY AMERICAN ROOMS: 1650-1858, edited by Russell Hawes Kettell. Tour of 12 rooms, each representative of a different era in American history and each furnished, decorated, designed and occupied in the style of the era. 72 plans and elevations, 8-page color section, etc., show fabrics, wall papers, arrangements, etc. Full descriptive text. xvii + 200pp. of text. 8⅜ x 11¼.
21633-0 Paperbound $5.00

THE FITZWILLIAM VIRGINAL BOOK, edited by J. Fuller Maitland and W. B. Squire. Full modern printing of famous early 17th-century ms. volume of 300 works by Morley, Byrd, Bull, Gibbons, etc. For piano or other modern keyboard instrument; easy to read format. xxxvi + 938pp. 8⅜ x 11.
21068-5, 21069-3 Two volumes, Paperbound $10.00

KEYBOARD MUSIC, Johann Sebastian Bach. Bach Gesellschaft edition. A rich selection of Bach's masterpieces for the harpsichord: the six English Suites, six French Suites, the six Partitas (Clavierübung part I), the Goldberg Variations (Clavierübung part IV), the fifteen Two-Part Inventions and the fifteen Three-Part Sinfonias. Clearly reproduced on large sheets with ample margins; eminently playable. vi + 312pp. 8⅛ x 11. 22360-4 Paperbound $5.00

THE MUSIC OF BACH: AN INTRODUCTION, Charles Sanford Terry. A fine, nontechnical introduction to Bach's music, both instrumental and vocal. Covers organ music, chamber music, passion music, other types. Analyzes themes, developments, innovations. x + 114pp. 21075-8 Paperbound $1.50

BEETHOVEN AND HIS NINE SYMPHONIES, Sir George Grove. Noted British musicologist provides best history, analysis, commentary on symphonies. Very thorough, rigorously accurate; necessary to both advanced student and amateur music lover. 436 musical passages. vii + 407 pp. 20334-4 Paperbound $2.75

JOHANN SEBASTIAN BACH, Philipp Spitta. One of the great classics of musicology, this definitive analysis of Bach's music (and life) has never been surpassed. Lucid, nontechnical analyses of hundreds of pieces (30 pages devoted to St. Matthew Passion, 26 to B Minor Mass). Also includes major analysis of 18th-century music. 450 musical examples. 40-page musical supplement. Total of xx + 1799pp.
(EUK) 22278-0, 22279-9 Two volumes, Clothbound $17.50

MOZART AND HIS PIANO CONCERTOS, Cuthbert Girdlestone. The only full-length study of an important area of Mozart's creativity. Provides detailed analyses of all 23 concertos, traces inspirational sources. 417 musical examples. Second edition. 509pp.
21271-8 Paperbound $3.50

THE PERFECT WAGNERITE: A COMMENTARY ON THE NIBLUNG'S RING, George Bernard Shaw. Brilliant and still relevant criticism in remarkable essays on Wagner's Ring cycle, Shaw's ideas on political and social ideology behind the plots, role of Leitmotifs, vocal requisites, etc. Prefaces. xxi + 136pp.
(USO) 21707-8 Paperbound $1.75

DON GIOVANNI, W. A. Mozart. Complete libretto, modern English translation; biographies of composer and librettist; accounts of early performances and critical reaction. Lavishly illustrated. All the material you need to understand and appreciate this great work. Dover Opera Guide and Libretto Series; translated and introduced by Ellen Bleiler. 92 illustrations. 209pp.
21134-7 Paperbound $2.00

BASIC ELECTRICITY, U. S. Bureau of Naval Personel. Originally a training course, best non-technical coverage of basic theory of electricity and its applications. Fundamental concepts, batteries, circuits, conductors and wiring techniques, AC and DC, inductance and capacitance, generators, motors, transformers, magnetic amplifiers, synchros, servomechanisms, etc. Also covers blue-prints, electrical diagrams, etc. Many questions, with answers. 349 illustrations. x + 448pp. 6½ x 9¼.
20973-3 Paperbound $3.50

REPRODUCTION OF SOUND, Edgar Villchur. Thorough coverage for laymen of high fidelity systems, reproducing systems in general, needles, amplifiers, preamps, loudspeakers, feedback, explaining physical background. "A rare talent for making technicalities vividly comprehensible," R. Darrell, *High Fidelity.* 69 figures. iv + 92pp.
21515-6 Paperbound $1.35

HEAR ME TALKIN' TO YA: THE STORY OF JAZZ AS TOLD BY THE MEN WHO MADE IT, Nat Shapiro and Nat Hentoff. Louis Armstrong, Fats Waller, Jo Jones, Clarence Williams, Billy Holiday, Duke Ellington, Jelly Roll Morton and dozens of other jazz greats tell how it was in Chicago's South Side, New Orleans, depression Harlem and the modern West Coast as jazz was born and grew. xvi + 429pp.
21726-4 Paperbound $3.00

FABLES OF AESOP, translated by Sir Roger L'Estrange. A reproduction of the very rare 1931 Paris edition; a selection of the most interesting fables, together with 50 imaginative drawings by Alexander Calder. v + 128pp. 6½x9¼.
21780-9 Paperbound $1.50

AGAINST THE GRAIN (A REBOURS), Joris K. Huysmans. Filled with weird images, evidences of a bizarre imagination, exotic experiments with hallucinatory drugs, rich tastes and smells and the diversions of its sybarite hero Duc Jean des Esseintes, this classic novel pushed 19th-century literary decadence to its limits. Full unabridged edition. Do not confuse this with abridged editions generally sold. Introduction by Havelock Ellis. xlix + 206pp. 22190-3 Paperbound $2.50

VARIORUM SHAKESPEARE: HAMLET. Edited by Horace H. Furness; a landmark of American scholarship. Exhaustive footnotes and appendices treat all doubtful words and phrases, as well as suggested critical emendations throughout the play's history. First volume contains editor's own text, collated with all Quartos and Folios. Second volume contains full first Quarto, translations of Shakespeare's sources (Belleforest, and Saxo Grammaticus), Der Bestrafte Brudermord, and many essays on critical and historical points of interest by major authorities of past and present. Includes details of staging and costuming over the years. By far the best edition available for serious students of Shakespeare. Total of xx + 905pp.
21004-9, 21005-7, 2 volumes, Paperbound $7.00

A LIFE OF WILLIAM SHAKESPEARE, Sir Sidney Lee. This is the standard life of Shakespeare, summarizing everything known about Shakespeare and his plays. Incredibly rich in material, broad in coverage, clear and judicious, it has served thousands as the best introduction to Shakespeare. 1931 edition. 9 plates. xxix + 792pp. 21967-4 Paperbound $3.75

MASTERS OF THE DRAMA, John Gassner. Most comprehensive history of the drama in print, covering every tradition from Greeks to modern Europe and America, including India, Far East, etc. Covers more than 800 dramatists, 2000 plays, with biographical material, plot summaries, theatre history, criticism, etc. "Best of its kind in English," New Republic. 77 illustrations. xxii + 890pp.
20100-7 Clothbound $10.00

THE EVOLUTION OF THE ENGLISH LANGUAGE, George McKnight. The growth of English, from the 14th century to the present. Unusual, non-technical account presents basic information in very interesting form: sound shifts, change in grammar and syntax, vocabulary growth, similar topics. Abundantly illustrated with quotations. Formerly Modern English in the Making. xii + 590pp.
21932-1 Paperbound $3.50

AN ETYMOLOGICAL DICTIONARY OF MODERN ENGLISH, Ernest Weekley. Fullest, richest work of its sort, by foremost British lexicographer. Detailed word histories, including many colloquial and archaic words; extensive quotations. Do not confuse this with the Concise Etymological Dictionary, which is much abridged. Total of xxvii + 830pp. 6½ x 9¼.
21873-2, 21874-0 Two volumes, Paperbound $7.90

FLATLAND: A ROMANCE OF MANY DIMENSIONS, E. A. Abbott. Classic of science-fiction explores ramifications of life in a two-dimensional world, and what happens when a three-dimensional being intrudes. Amusing reading, but also useful as introduction to thought about hyperspace. Introduction by Banesh Hoffmann. 16 illustrations. xx + 103pp. 20001-9 Paperbound $1.00

POEMS OF ANNE BRADSTREET, edited with an introduction by Robert Hutchinson. A new selection of poems by America's first poet and perhaps the first significant woman poet in the English language. 48 poems display her development in works of considerable variety—love poems, domestic poems, religious meditations, formal elegies, "quaternions," etc. Notes, bibliography. viii + 222pp.

22160-1 Paperbound $2.50

THREE GOTHIC NOVELS: THE CASTLE OF OTRANTO BY HORACE WALPOLE; VATHEK BY WILLIAM BECKFORD; THE VAMPYRE BY JOHN POLIDORI, WITH FRAGMENT OF A NOVEL BY LORD BYRON, edited by E. F. Bleiler. The first Gothic novel, by Walpole; the finest Oriental tale in English, by Beckford; powerful Romantic supernatural story in versions by Polidori and Byron. All extremely important in history of literature; all still exciting, packed with supernatural thrills, ghosts, haunted castles, magic, etc. xl + 291pp.

21232-7 Paperbound $2.50

THE BEST TALES OF HOFFMANN, E. T. A. Hoffmann. 10 of Hoffmann's most important stories, in modern re-editings of standard translations: Nutcracker and the King of Mice, Signor Formica, Automata, The Sandman, Rath Krespel, The Golden Flowerpot, Master Martin the Cooper, The Mines of Falun, The King's Betrothed, A New Year's Eve Adventure. 7 illustrations by Hoffmann. Edited by E. F. Bleiler. xxxix + 419pp.

21793-0 Paperbound $3.00

GHOST AND HORROR STORIES OF AMBROSE BIERCE, Ambrose Bierce. 23 strikingly modern stories of the horrors latent in the human mind: The Eyes of the Panther, The Damned Thing, An Occurrence at Owl Creek Bridge, An Inhabitant of Carcosa, etc., plus the dream-essay, Visions of the Night. Edited by E. F. Bleiler. xxii + 199pp.

20767-6 Paperbound $1.50

BEST GHOST STORIES OF J. S. LEFANU, J. Sheridan LeFanu. Finest stories by Victorian master often considered greatest supernatural writer of all. Carmilla, Green Tea, The Haunted Baronet, The Familiar, and 12 others. Most never before available in the U. S. A. Edited by E. F. Bleiler. 8 illustrations from Victorian publications. xvii + 467pp.

20415-4 Paperbound $3.00

MATHEMATICAL FOUNDATIONS OF INFORMATION THEORY, A. I. Khinchin. Comprehensive introduction to work of Shannon, McMillan, Feinstein and Khinchin, placing these investigations on a rigorous mathematical basis. Covers entropy concept in probability theory, uniqueness theorem, Shannon's inequality, ergodic sources, the E property, martingale concept, noise, Feinstein's fundamental lemma, Shanon's first and second theorems. Translated by R. A. Silverman and M. D. Friedman. iii + 120pp.

60434-9 Paperbound $2.00

SEVEN SCIENCE FICTION NOVELS, H. G. Wells. The standard collection of the great novels. Complete, unabridged. *First Men in the Moon, Island of Dr. Moreau, War of the Worlds, Food of the Gods, Invisible Man, Time Machine, In the Days of the Comet.* Not only science fiction fans, but every educated person owes it to himself to read these novels. 1015pp. (USO) 20264-X Clothbound $6.00

LAST AND FIRST MEN AND STAR MAKER, TWO SCIENCE FICTION NOVELS, Olaf Stapledon. Greatest future histories in science fiction. In the first, human intelligence is the "hero," through strange paths of evolution, interplanetary invasions, incredible technologies, near extinctions and reemergences. Star Maker describes the quest of a band of star rovers for intelligence itself, through time and space: weird inhuman civilizations, crustacean minds, symbiotic worlds, etc. Complete, unabridged. v + 438pp. (USO) 21962-3 Paperbound $2.50

THREE PROPHETIC NOVELS, H. G. WELLS. Stages of a consistently planned future for mankind. *When the Sleeper Wakes,* and *A Story of the Days to Come,* anticipate *Brave New World* and *1984,* in the 21st Century; *The Time Machine,* only complete version in print, shows farther future and the end of mankind. All show Wells's greatest gifts as storyteller and novelist. Edited by E. F. Bleiler. x + 335pp. (USO) 20605-X Paperbound $2.50

THE DEVIL'S DICTIONARY, Ambrose Bierce. America's own Oscar Wilde— Ambrose Bierce—offers his barbed iconoclastic wisdom in over 1,000 definitions hailed by H. L. Mencken as "some of the most gorgeous witticisms in the English language." 145pp. 20487-1 Paperbound $1.25

MAX AND MORITZ, Wilhelm Busch. Great children's classic, father of comic strip, of two bad boys, Max and Moritz. Also Ker and Plunk (Plisch und Plumm), Cat and Mouse, Deceitful Henry, Ice-Peter, The Boy and the Pipe, and five other pieces. Original German, with English translation. Edited by H. Arthur Klein; translations by various hands and H. Arthur Klein. vi + 216pp. 20181-3 Paperbound $2.00

PIGS IS PIGS AND OTHER FAVORITES, Ellis Parker Butler. The title story is one of the best humor short stories, as Mike Flannery obfuscates biology and English. Also included, That Pup of Murchison's, The Great American Pie Company, and Perkins of Portland. 14 illustrations. v + 109pp. 21532-6 Paperbound $1.25

THE PETERKIN PAPERS, Lucretia P. Hale. It takes genius to be as stupidly mad as the Peterkins, as they decide to become wise, celebrate the "Fourth," keep a cow, and otherwise strain the resources of the Lady from Philadelphia. Basic book of American humor. 153 illustrations. 219pp. 20794-3 Paperbound $2.00

PERRAULT'S FAIRY TALES, translated by A. E. Johnson and S. R. Littlewood, with 34 full-page illustrations by Gustave Doré. All the original Perrault stories— Cinderella, Sleeping Beauty, Bluebeard, Little Red Riding Hood, Puss in Boots, Tom Thumb, etc.—with their witty verse morals and the magnificent illustrations of Doré. One of the five or six great books of European fairy tales. viii + 117pp. 8⅛ x 11. 22311-6 Paperbound $2.00

OLD HUNGARIAN FAIRY TALES, Baroness Orczy. Favorites translated and adapted by author of the *Scarlet Pimpernel.* Eight fairy tales include "The Suitors of Princess Fire-Fly," "The Twin Hunchbacks," "Mr. Cuttlefish's Love Story," and "The Enchanted Cat." This little volume of magic and adventure will captivate children as it has for generations. 90 drawings by Montagu Barstow. 96pp.
(USO) 22293-4 Paperbound $1.95

THE RED FAIRY BOOK, Andrew Lang. Lang's color fairy books have long been children's favorites. This volume includes Rapunzel, Jack and the Bean-stalk and 35 other stories, familiar and unfamiliar. 4 plates, 93 illustrations x + 367pp.
21673-X Paperbound $2.50

THE BLUE FAIRY BOOK, Andrew Lang. Lang's tales come from all countries and all times. Here are 37 tales from Grimm, the Arabian Nights, Greek Mythology, and other fascinating sources. 8 plates, 130 illustrations. xi + 390pp.
21437-0 Paperbound $2.50

HOUSEHOLD STORIES BY THE BROTHERS GRIMM. Classic English-language edition of the well-known tales — Rumpelstiltskin, Snow White, Hansel and Gretel, The Twelve Brothers, Faithful John, Rapunzel, Tom Thumb (52 stories in all). Translated into simple, straightforward English by Lucy Crane. Ornamented with headpieces, vignettes, elaborate decorative initials and a dozen full-page illustrations by Walter Crane. x + 269pp.
21080-4 Paperbound **$2.00**

THE MERRY ADVENTURES OF ROBIN HOOD, Howard Pyle. The finest modern versions of the traditional ballads and tales about the great English outlaw. Howard Pyle's complete prose version, with every word, every illustration of the first edition. Do not confuse this facsimile of the original (1883) with modern editions that change text or illustrations. 23 plates plus many page decorations. xxii + 296pp.
22043-5 Paperbound $2.50

THE STORY OF KING ARTHUR AND HIS KNIGHTS, Howard Pyle. The finest children's version of the life of King Arthur; brilliantly retold by Pyle, with 48 of his most imaginative illustrations. xviii + 313pp. 6⅛ x 9¼.
21445-1 Paperbound $2.50

THE WONDERFUL WIZARD OF OZ, L. Frank Baum. America's finest children's book in facsimile of first edition with all Denslow illustrations in full color. The edition a child should have. Introduction by Martin Gardner. 23 color plates, scores of drawings. iv + 267pp.
20691-2 Paperbound $2.50

THE MARVELOUS LAND OF OZ, L. Frank Baum. The second Oz book, every bit as imaginative as the Wizard. The hero is a boy named Tip, but the Scarecrow and the Tin Woodman are back, as is the Oz magic. 16 color plates, 120 drawings by John R. Neill. 287pp.
20692-0 Paperbound $2.50

THE MAGICAL MONARCH OF MO, L. Frank Baum. Remarkable adventures in a land even stranger than Oz. The best of Baum's books not in the Oz series. 15 color plates and dozens of drawings by Frank Verbeck. xviii + 237pp.
21892-9 Paperbound $2.25

THE BAD CHILD'S BOOK OF BEASTS, MORE BEASTS FOR WORSE CHILDREN, A MORAL ALPHABET, Hilaire Belloc. Three complete humor classics in one volume. Be kind to the frog, and do not call him names . . . and 28 other whimsical animals. Familiar favorites and some not so well known. Illustrated by Basil Blackwell. 156pp.
(USO) 20749-8 Paperbound $1.50

EAST O' THE SUN AND WEST O' THE MOON, George W. Dasent. Considered the best of all translations of these Norwegian folk tales, this collection has been enjoyed by generations of children (and folklorists too). Includes True and Untrue, Why the Sea is Salt, East O' the Sun and West O' the Moon, Why the Bear is Stumpy-Tailed, Boots and the Troll, The Cock and the Hen, Rich Peter the Pedlar, and 52 more. The only edition with all 59 tales. 77 illustrations by Erik Werenskiold and Theodor Kittelsen. xv + 418pp. 22521-6 Paperbound $3.50

GOOPS AND HOW TO BE THEM, Gelett Burgess. Classic of tongue-in-cheek humor, masquerading as etiquette book. 87 verses, twice as many cartoons, show mischievous Goops as they demonstrate to children virtues of table manners, neatness, courtesy, etc. Favorite for generations. viii + 88pp. 6½ x 9¼.
22233-0 Paperbound $1.25

ALICE'S ADVENTURES UNDER GROUND, Lewis Carroll. The first version, quite different from the final *Alice in Wonderland,* printed out by Carroll himself with his own illustrations. Complete facsimile of the "million dollar" manuscript Carroll gave to Alice Liddell in 1864. Introduction by Martin Gardner. viii + 96pp. Title and dedication pages in color. 21482-6 Paperbound $1.25

THE BROWNIES, THEIR BOOK, Palmer Cox. Small as mice, cunning as foxes, exuberant and full of mischief, the Brownies go to the zoo, toy shop, seashore, circus, etc., in 24 verse adventures and 266 illustrations. Long a favorite, since their first appearance in St. Nicholas Magazine. xi + 144pp. 6⅝ x 9¼.
21265-3 Paperbound $1.75

SONGS OF CHILDHOOD, Walter De La Mare. Published (under the pseudonym Walter Ramal) when De La Mare was only 29, this charming collection has long been a favorite children's book. A facsimile of the first edition in paper, the 47 poems capture the simplicity of the nursery rhyme and the ballad, including such lyrics as I Met Eve, Tartary, The Silver Penny. vii + 106pp. (USO) 21972-0 Paperbound
$1.25

THE COMPLETE NONSENSE OF EDWARD LEAR, Edward Lear. The finest 19th-century humorist-cartoonist in full: all nonsense limericks, zany alphabets, Owl and Pussycat, songs, nonsense botany, and more than 500 illustrations by Lear himself. Edited by Holbrook Jackson. xxix + 287pp. (USO) 20167-8 Paperbound $2.00

BILLY WHISKERS: THE AUTOBIOGRAPHY OF A GOAT, Frances Trego Montgomery. A favorite of children since the early 20th century, here are the escapades of that rambunctious, irresistible and mischievous goat—Billy Whiskers. Much in the spirit of *Peck's Bad Boy,* this is a book that children never tire of reading or hearing. All the original familiar illustrations by W. H. Fry are included: 6 color plates, 18 black and white drawings. 159pp. 22345-0 Paperbound $2.00

MOTHER GOOSE MELODIES. Faithful republication of the fabulously rare Munroe and Francis "copyright 1833" Boston edition—the most important Mother Goose collection, usually referred to as the "original." Familiar rhymes plus many rare ones, with wonderful old woodcut illustrations. Edited by E. F. Bleiler. 128pp. 4½ x 6⅜. 22577-1 Paperbound $1.00

TWO LITTLE SAVAGES; BEING THE ADVENTURES OF TWO BOYS WHO LIVED AS INDIANS AND WHAT THEY LEARNED, Ernest Thompson Seton. Great classic of nature and boyhood provides a vast range of woodlore in most palatable form, a genuinely entertaining story. Two farm boys build a teepee in woods and live in it for a month, working out Indian solutions to living problems, star lore, birds and animals, plants, etc. 293 illustrations. vii + 286pp.

20985-7 Paperbound $2.50

PETER PIPER'S PRACTICAL PRINCIPLES OF PLAIN & PERFECT PRONUNCIATION. Alliterative jingles and tongue-twisters of surprising charm, that made their first appearance in America about 1830. Republished in full with the spirited woodcut illustrations from this earliest American edition. 32pp. 4½ x 6⅜.

22560-7 Paperbound $1.00

SCIENCE EXPERIMENTS AND AMUSEMENTS FOR CHILDREN, Charles Vivian. 73 easy experiments, requiring only materials found at home or easily available, such as candles, coins, steel wool, etc.; illustrate basic phenomena like vacuum, simple chemical reaction, etc. All safe. Modern, well-planned. Formerly *Science Games for Children*. 102 photos, numerous drawings. 96pp. 6⅛ x 9¼.

21856-2 Paperbound $1.25

AN INTRODUCTION TO CHESS MOVES AND TACTICS SIMPLY EXPLAINED, Leonard Barden. Informal intermediate introduction, quite strong in explaining reasons for moves. Covers basic material, tactics, important openings, traps, positional play in middle game, end game. Attempts to isolate patterns and recurrent configurations. Formerly *Chess*. 58 figures. 102pp. (USO) 21210-6 Paperbound $1.25

LASKER'S MANUAL OF CHESS, Dr. Emanuel Lasker. Lasker was not only one of the five great World Champions, he was also one of the ablest expositors, theorists, and analysts. In many ways, his Manual, permeated with his philosophy of battle, filled with keen insights, is one of the greatest works ever written on chess. Filled with analyzed games by the great players. A single-volume library that will profit almost any chess player, beginner or master. 308 diagrams. xli x 349pp.

20640-8 Paperbound $2.75

THE MASTER BOOK OF MATHEMATICAL RECREATIONS, Fred Schuh. In opinion of many the finest work ever prepared on mathematical puzzles, stunts, recreations; exhaustively thorough explanations of mathematics involved, analysis of effects, citation of puzzles and games. Mathematics involved is elementary. Translated bv F. Göbel. 194 figures. xxiv + 430pp. 22134-2 Paperbound $3.50

MATHEMATICS, MAGIC AND MYSTERY, Martin Gardner. Puzzle editor for Scientific American explains mathematics behind various mystifying tricks: card tricks, stage "mind reading," coin and match tricks, counting out games, geometric dissections, etc. Probability sets, theory of numbers clearly explained. Also provides more than 400 tricks, guaranteed to work, that you can do. 135 illustrations. xii + 176pp.

20335-2 Paperbound $1.75

CATALOGUE OF DOVER BOOKS

MATHEMATICAL PUZZLES FOR BEGINNERS AND ENTHUSIASTS, Geoffrey Mott-Smith. 189 puzzles from easy to difficult—involving arithmetic, logic, algebra, properties of digits, probability, etc.—for enjoyment and mental stimulus. Explanation of mathematical principles behind the puzzles. 135 illustrations. viii + 248pp.
20198-8 Paperbound $1.75

PAPER FOLDING FOR BEGINNERS, William D. Murray and Francis J. Rigney. Easiest book on the market, clearest instructions on making interesting, beautiful origami. Sail boats, cups, roosters, frogs that move legs, bonbon boxes, standing birds, etc. 40 projects; more than 275 diagrams and photographs. 94pp.
20713-7 Paperbound $1.00

TRICKS AND GAMES ON THE POOL TABLE, Fred Herrmann. 79 tricks and games—some solitaires, some for two or more players, some competitive games—to entertain you between formal games. Mystifying shots and throws, unusual caroms, tricks involving such props as cork, coins, a hat, etc. Formerly *Fun on the Pool Table*. 77 figures. 95pp.
21814-7 Paperbound $1.25

HAND SHADOWS TO BE THROWN UPON THE WALL: A SERIES OF NOVEL AND AMUSING FIGURES FORMED BY THE HAND, Henry Bursill. Delightful picturebook from great-grandfather's day shows how to make 18 different hand shadows: a bird that flies, duck that quacks, dog that wags his tail, camel, goose, deer, boy, turtle, etc. Only book of its sort. vi + 33pp. 6½ x 9¼. 21779-5 Paperbound $1.00

WHITTLING AND WOODCARVING, E. J. Tangerman. 18th printing of best book on market. "If you can cut a potato you can carve" toys and puzzles, chains, chessmen, caricatures, masks, frames, woodcut blocks, surface patterns, much more. Information on tools, woods, techniques. Also goes into serious wood sculpture from Middle Ages to present, East and West. 464 photos, figures. x + 293pp.
20965-2 Paperbound $2.00

HISTORY OF PHILOSOPHY, Julián Marias. Possibly the clearest, most easily followed, best planned, most useful one-volume history of philosophy on the market; neither skimpy nor overfull. Full details on system of every major philosopher and dozens of less important thinkers from pre-Socratics up to Existentialism and later. Strong on many European figures usually omitted. Has gone through dozens of editions in Europe. 1966 edition, translated by Stanley Appelbaum and Clarence Strowbridge. xviii + 505pp. 21739-6 Paperbound $3.50

YOGA: A SCIENTIFIC EVALUATION, Kovoor T. Behanan. Scientific but non-technical study of physiological results of yoga exercises; done under auspices of Yale U. Relations to Indian thought, to psychoanalysis, etc. 16 photos. xxiii + 270pp.
20505-3 Paperbound $2.50

Prices subject to change without notice.
Available at your book dealer or write for free catalogue to Dept. GI, Dover Publications, Inc., 180 Varick St., N. Y., N. Y. 10014. Dover publishes more than 150 books each year on science, elementary and advanced mathematics, biology, music, art, literary history, social sciences and other areas.